Great Geologists

By

M. D. Simmons

HALLIBURTON

First published 2018

Halliburton
97 Jubilee Avenue
Milton Park
Abingdon
OX14 4RW
United Kingdom

www.landmark.solutions/NeftexInsights

Copyright © Halliburton
All rights reserved. This does not cover photographs and other illustrations provided by third parties, who retain copyright of their images; reproduction permission for these images must be sought from the copyright holders.

The right of Michael Derek Simmons to be identified as the Author of this work has been asserted in accordance with the Copyrights, Designs and Patents Act 1988.

ISBN 978-1-9160054-1-9 (print)
ISBN 978-1-9160054-0-2 (ebook) (https://joom.ag/ggLa)

British Library Cataloguing in Publication Data
A catalogue record for this book is available from the British Library

Printed and bound by CPI Group (UK) Ltd, Croydon, CR0 4YY, U.K.

Disclaimer

This book is a compilation of articles written by Mike Simmons and previously published in Neftex® Exploration Insights magazine between September 2015 and November 2018. This compilation was made in November 2018. Each article is a synthesis based upon published data and information, and derived knowledge created within Halliburton. Unless explicitly stated otherwise, no proprietary client data has been used in its preparation. If client data has been used, permission will have been obtained and is acknowledged. Reproduction of any copyrighted image is with the permission of the copyright holder and is acknowledged. The opinions found in the articles may not necessarily reflect the views and/or opinions of Halliburton Energy Services, Inc. and its affiliates including but not limited to Landmark Graphics Corporation.

Great Geologists: An Introduction

Prof. Mike Simmons
HALLIBURTON TECHNOLOGY FELLOW

Why a Compilation of Great Geologists?

To the best of my knowledge, no compilation exists that provides succinct biographic summaries of Great Geologists, notwithstanding some excellent reviews of the history of geological thinking that focus on the development of geological theories (see reference list). The history of geoscience is marked by the work of exemplary scientists, who through their endeavours, changed the way we think about the Earth, its history, processes and resources. Some made huge intuitive leaps, recognising, for example, the immensity of geological time or the mobility of the continents. Others described rocks, minerals and fossils in the field, or laboratory, and provided vital data that allowed theories to develop. Others still embraced new technologies, such as geophysics, that enabled what cannot be observed directly to be interpreted. Many led colourful lives or overcame adverse circumstances. These seem like people worth knowing more about, not least, for the inspiration they provide.

Geoscience is becoming increasingly specialized, numerical and driven by advances in computing, data science and by other technological innovations across a broad front. In these exciting times, it is easy to forget the founders of geology, the giants on whose shoulders we stand today. The history of geology is no longer taught at many universities. Many early career geologists have little idea who the Greats are or what their key accomplishments were. This is a pity because there is much to learn from geoscience leaders past and present, not least their approaches to scientific problems and their persistent endeavours, often in spite of difficulties or controversy. Their stories are inspirational and, if nothing more, give us role models that we can aspire towards.

What Makes a Geologist Great?

Any selection of Great Geologists must be a personal one, and I make no apologies for the fact that this is the case here. Once one has considered the truly astonishing insights of the founders of our science, James

Hutton and Sir Charles Lyell for example, selecting other Greats becomes somewhat subjective. Is the geological mapping of a remote corner of the globe as important an achievement as discovering a new group of minerals? What greatness should be attributed to brilliant teachers and communicators of geoscience? It seems sensible to classify as "Great" those who developed important new theories and changed the way we think about the Earth and its history. However, we should also seek to include some of those who tirelessly gathered data, normally in the field (the natural home of the geologist) or the laboratory, that made the giant leaps in geological insight possible.

My background is in stratigraphy, micropalaeontology and regional petroleum geology, so my choices are a little coloured by that. I have also consciously tried to recognise the national and gender diversity of the Greats in geological research. Nonetheless, if there are a good number of British gentlemen scientists mentioned, that simply reflects the historical realities influencing how and where geology unfolded. The British Isles contain a great variety of geology in a small area. This was initially researched by men of means at a time when Britain was a powerhouse for economic development and learning.

I have not, however, tried to make this into an exhaustive selection and some readers may be disappointed that I did not select their personal hero. My intent has been to provide a broad coverage of the architects of revolutions in geoscience and those who assisted that process by contributing exceptional work.

The history of science is no place for icons. Even the indisputably Great have their theories nit-picked by those scientists that follow them. Great Geologists, as with all scientists who have pushed the boundaries of knowledge, were, and are, not always right in their opinions. For example, before the advent of plate tectonics, many geologists, including Greats like Eduard Suess, sought to explain mountain building in the context of a contracting Earth. Nonetheless, such errors do not preclude them from being considered as Great — it is their whole body of work and adding to the progression of geoscience that marks them so. Geology, along with other sciences, is self-correcting. Errors in the effort to elucidate that which can never truly be known, unless a time machine is invented, can be excused if the overall effect is to move the science forward.

Of more concern for inclusion are those who hold on to outdated theories despite the mass of evidence that disproves them. Having said this, geologists, as in many other areas of science, have to work within the observational and technological limits of their time. Many geologists can be considered as Great simply because they went out into the field and gathered data where previously none existed. Without new data, fresh observation, or innovative ways of analysing old data, there can be no progress in any science.

Thinking About the Earth

The history of geological thinking is a long one, with scholars in both ancient Greece and Rome contemplating the history of the Earth and how that related to the rocks beneath their feet. In the 5th century B.C., Xanthus of Lydia saw shell shapes in rocks now located far from the coast and concluded that these regions must have once been submerged beneath the sea. Centuries later, Leonardo da Vinci drew similar conclusions.

Nonetheless, it was not until the mid-17th century and the arrival of the Dane Nicolas Steno in Late Renaissance Florence that modern geological thinking can be said to have started. Steno's observations, during his brief dalliance with geology, were seemingly simple by modern standards: in a normal succession of rocks, the oldest are at the bottom and the youngest at the top; sedimentary rocks are laid down horizontally; if they are not horizontal, then they have been folded or faulted; while fossils are the preserved remains of ancient creatures. Yet, these notions suggested that the rock record and its fossil content represented a chronology, effectively a book of Earth history, waiting to be read.

It was not until over 100 years later that the book of Earth history began to be read in earnest. During the Age of Enlightenment, the Scottish intellectual James Hutton argued that from a consideration of the processes creating rocks and their subsequent deformation, the age of the Earth had to be, by human standards, immense ("the abyss of time" as described by Hutton's friend and fellow intellectual, John Playfair). How immense was uncertain, but certainly much older than might be determined from a literal interpretation of the Bible, or other religious texts.

Geology as a stand-alone subject was born in the late 18th century with the work of Hutton and others. Hutton was not the "Father of Geology" as he is sometimes portrayed, but he was an important catalyst in developing inductive thinking about the age of the Earth and geological processes. Others such as Abraham Gottlob Werner, Georges Louis Leclerc, Comte de Buffon and Peter Pallas were contemplating similar issues in Germany, France and Russia, respectively. Around the time of Hutton, geology as a term with its current meaning was introduced, by the Geneva-based naturalist Jean-André Deluc in 1778, although it had been used with a broader meaning (including the study of plants and animals) since the 15th century (as the Latin word *geologia*).

Werner and Hutton were on opposite sides of the controversy that existed between "Neptunists" and "Plutonists", which occupied geological, and indeed popular, thinking in the late 18th century. Werner had promoted the notion that all rocks, including granites and basalts, were either deposited or precipitated out of water ("Neptunism"), whilst Hutton favoured the plutonic view that granites and basalts were the products of heat within the Earth creating molten magma ("Plutonism"). His observation of cross-cutting intrusions demonstrated this. By the beginning of the 19th century, Neptunism as an explanation for crystalline rocks, such as granite, was effectively no longer in vogue. Instead, rock classifications concentrated on the concepts we now know as igneous, metamorphic and sedimentary.

The next major step in the history of geology was to determine that the fossil content of sedimentary rocks could be used as a key to understanding that a given rock unit could be associated with a specific period of Earth history. This allowed correlation to other rocks deposited during the same period — the science of stratigraphy was born. Recognition of distinct strata permitted the mapping of these layers as they occurred at the Earth's surface and, equally importantly, enabled a prediction to be made of what might lie *below the surface*. William Smith in England and Georges Cuvier in France pioneered this thinking at the end of the 18th century into the beginning of the 19th century.

Geological research focused on two distinct activities for the first half of the 19th century. There were those concerned with the description and classification of rocks, minerals and fossils and, most notably, the subdivision of Earth history. British geologists such as Sir Roderick Murchison and Adam Sedgewick were at the forefront of this campaign, with European counterparts such as Alcide d'Orbigny not far behind.

Other researchers were concerned with the geological processes operating on and within the Earth, and how these processes may have operated in the geological past. In other words, how rocks came to be formed and subsequently deformed. Foremost amongst these was Sir Charles Lyell, who considered himself on a crusade to make geology scientific. Observations led to theories about how geological processes operated. In Lyell's view, these processes were gradualistic — a steady state Earth in which geological processes were the same in the past ("Uniformitarianism", following on from ideas earlier expressed by Hutton). By contrast, many geologists in continental Europe favoured the theory promoted by Georges Cuvier that the Earth had experienced a more eventful past, with catastrophes punctuating Earth history, these events being associated with tectonics, extinctions and major changes in deposition ("Catastrophism"). The debate of the importance of Uniformitarianism versus Catastrophism continued throughout much of the 19th century and persists in some circles even today; although most geologists are now happy to accept that Earth history is a response to a combination of both gradual and sudden processes.

However, many 19th century geologists (as today) were both describers/classifiers and interpreters, attempting to add colour to the pages of Earth history by envisaging past worlds. What did a Jurassic Earth look like and what creatures inhabited it? Which geological processes were operating to leave us with the rock record we see today? Such intriguing questions are equally valid nowadays and engage the imagination of most geologists to a greater or lesser extent, even if their focus is often on the fine detail. The romance of imagining our past Earth is something that still draws students to study geology and requires both an understanding of geological classification and of geological processes.

By the second half of the 19th century, much of the basic classification work had been completed (although this continues to the present day in order to provide ever-increasing precision) and greater numbers of geological scholars were focused on interpreting the rocks they studied in terms of the processes responsible for their creation and deformation. Such studies ranged from the small-scale, for example, Henry Sorby and his interpretation of rocks in microscopic thin-sections, to the large scale, such as Eduard Suess and his interpretation of the formation of mountain belts. Although Western Europe continued to be a hub for geological research, American researchers, such as James Dana and Louis Agassiz, were now also making important contributions.

Geologists, however, were still faced with the perplexing conundrum — how old was the Earth? It was widely accepted to be millions of years in duration, but exactly how many remained an unknown. The discovery of radioactivity, as the 19th century passed into the 20th century, provided the breakthrough. Radioactive decay of elements present in certain rocks could be measured and interpreted in terms of absolute age. At last, there was a clock of Earth history!

The undisputed pioneer of this research was the Great British geologist Arthur Holmes. Thanks to Holmes, and those who followed him, geologists could use an understanding of the age of the Earth and the duration of the chapters in Earth history to help elucidate the processes responsible for forming and deforming rocks, and explain modes of evolution represented by the fossil record. Knowledge of true geological time allowed thinking on a grand scale. Alfred Wegener was able to envisage continents drifting on the Earth's surface throughout geological time; whilst others, such as T.C. Chamberlin and Amadeus Grabau, began to recognise a rhythm to the Earth's sedimentary record. These notions, in turn, spurred the paradigms of plate tectonics and sequence stratigraphy in the second half of the 20th century.

The puzzles faced by geologists were now becoming increasingly difficult to be solved by a single researcher. The plate tectonics paradigm is to geology what evolutionary theory is to biology. It is very convenient to associate major scientific breakthroughs with a particular individual. Thus, Charles Darwin is often celebrated as the 'discoverer' of evolution. However, the discovery of the deep time in which geological processes operate, for example, cannot be said to be the work of James Hutton alone — he built on the notions expressed by many scholars of his time and before him.

Even though geologists are often asked who 'discovered' plate tectonics, the answer is that no single person can be said to have done so. Papers by Dan McKenzie and Bob Parker or by Jason Morgan can be cited as being the first to describe plate motions as translations and rotations on a sphere, but these built upon a long series of discoveries by many other researchers who worked on the bathymetry of the deep ocean, such as Mary Tharp, the nature of oceanic and continental crust, sea-floor spreading, transform faults, and convection within the interior of the Earth. Plate tectonics is arguably the last great geological discovery — the culmination and integration of understanding geological time and geological processes. Of course, new geological discoveries are made every day, but nothing (as yet!) can compare to the scale of the plate tectonics paradigm.

Geoscience is becoming increasingly collaborative. Integrated research requires multiple specialists to work together, each often from distant parts of the globe. Thus, the scientific societies operating within geology have played an important role as forums for the exchange of ideas and to bring scientists together. The Geological Society of London was founded in 1807 and there are now societies operating in almost every country and at an international level. UNESCO projects also encourage collaboration. Geology is now truly global in its outlook!

Peak Geoscience?

In the same way that people use the term "Peak Oil" to describe the time of peak oil supply to the market, the term "Peak Geoscience" has recently been heard. Briefly: has our development of geological science reached its zenith? Therefore, have we entered into a time of synthesis, the gathering of detail and of consolidation? Personally, I doubt this. Geological research still has many questions to answer and the application of data science is likely to reveal startling new insights.

Geology is becoming increasingly holistic and integrated with other sciences. Earth Systems Science integrates a variety of processes operating at the full range of timescales (from hours to millions of years) and spatial contexts (from local depositional processes to global tectonics), thereby providing insight into sediment supply from mountain source to sediment sink within a depositional basin. Palaeoclimate research, such as that carried out by Maureen Raymo, works in a similar manner, providing insights for modelling of future climate trends. Alongside such holistic approaches, it is likely that advances in technology and data science will transform geology by teasing out patterns in geological data that are beyond the capacity for easy recognition by humans.

Geological research continues across all branches of the subject and is expanding from the confines of the Earth to consider the geology of neighbouring planets and their moons. No doubt there will be more Great Geologists to emerge and be recognised over the coming years.

It is not easy to imagine what the geological breakthroughs of the future will be. Will we be able to predict the precise timing of earthquakes and volcanic eruptions? Will we develop techniques to directly date *any rock* accurately; thereby, improving correlation immeasurably? What seems certain is that technology will provide the catalyst.

The geologists working before the 20th century went out into the field and made observations that then led to theories of geological processes. During the 20th century, technological breakthroughs made a huge difference to geological thinking. For example, the advent of radiometric dating, the recognition of paleomagnetism, and the use of geophysical techniques and remote sensing are immense. That does not belie the importance of field work — there is no substitute for gathering data — but rather than armed simply with a hammer, hand-lens, compass-clinometer and paper notebook, the field geologist now has a wider variety and more sophisticated set of tools at their disposal, including drones and 3D imaging technology. Geoscience is far from being at its peak!

This Volume

As the reader may have already determined, this compilation describing Great Geologists is not a scholarly work. For such texts, the reader is encouraged to consult the exemplary writings of Martin Rudwick, David Oldroyd, Mott Greene and Henry Frankel, who have been amongst our foremost historians of geological research. Instead, this compilation is intended to provide the reader with brief portraits of some Great Geologists, outlining their main achievements and providing a little background colour about their lives. I was inspired by the geologists I have written about. It is hoped that readers will be similarly inspired, whilst gaining insight into how geological research is undertaken and what the results can tell us about our planet, its history and the processes operating on and within it.

Bibliography

A short list of references is given at end of each Great Geologist portrait. The following are works which I have turned to time and time again for information on the history of geology and its key players.
Frankel, H.R. 2012. *The Continental Drift Controversy.* Cambridge University Press, Volumes 1-4.
Gohau, G. 1991. *A History of Geology.* Rutgers University Press.
Greene, M.T. 1982. *Geology in the Nineteenth Century.* Cornell University Press.
Hallam, A. 1989. *Great Geologic Controversies.* Oxford University Press.
O'Hara, K.D.2018. *A Brief History of Geology.* Cambridge University Press.
Oldroyd, D. 1996. *Thinking About the Earth.* Harvard University Press.
Rudwick, M.J.S. 2007. *Bursting the Limits of Time.* University of Chicago Press.
Rudwick, M.J.S. 2014. *Earth's Deep History.* University of Chicago Press.

The key sources of information used in compiling the portrait of each Great Geologist are given within the relevent chapter. I am not ashamed to admit that Wikipedia has proven to be a valuable resource, although every effort has been made to check the information it mentions.

Acknowledgements

My colleague Rebecca Head has much improved this volume through her meticulous proof-reading and helpful modifications. Andy Hill of BP has been an enthusiastic reader and provided me with much useful advice, as have Roger Davies and Dave Casey. Liz Donnelly has formatted this book into an attractive layout. Mistakes of fact and interpretation are mine alone. Picture credits are at the end of this book. Halliburton is thanked for facilitating the publication of this book.

CONTENTS

NICOLAS STENO 12

ABRAHAM WERNER 14

JAMES HUTTON 18

WILLIAM SMITH 21

GEORGES CUVIER 24

MARY ANNING 28

WILLIAM BUCKLAND 30

ADAM SEDGWICK 33

SIR RODERICK MURCHISON 36

SIR CHARLES LYELL 39

ALCIDE D'ORBIGNY 42

CHARLES DARWIN 46

LOUIS AGASSIZ 51

JAMES DWIGHT DANA 55

CHARLES LAPWORTH 59

HENRY CLIFTON SORBY 64

EDUARD SUESS 68

T.C. CHAMBERLIN 71

ALEXANDER KARPINSKY 74

BEN PEACH & JOHN HORNE 77

ALFRED WEGENER 82

ARTHUR HOLMES 85

MILUTIN MILANKOVIĆ 89

AMADEUS GRABAU 94

WILLIAM JOCELYN ARKELL 97

MARIE THARP 100

ZIAD BEYDOUN 104

HARRY HESS 107

FRED VINE 111

JOHN TUZO WILSON 116

JANET WATSON 120

BRIAN HARLAND 123

DAN MCKENZIE 126

PETER VAIL 129

MAUREEN RAYMO 132

Nicolas Steno

One day in October 1666, two fishermen from the town of Livorno landed a huge shark they had caught off the coast of Tuscany. News of this reached Grand Duke Ferdinand II in Florence who ordered the head to be brought to the city to be dissected by a Danish anatomist attached to his court. The anatomist was Nicolaus Steno (also known by his native name Niels Stensen). The results of his study were to provide the sparks that ignited the sciences of paleontology and geology as we know them now.

Steno was born in 1638, the son of a Copenhagen goldsmith. At the age of 19, he entered the University of Copenhagen to study medicine. He then travelled to Holland and Paris where he honed his skills in anatomy and precise dissection.

During the mid-17th century, the understanding of how muscles, the heart and brain function was based on deductive reasoning rather than direct observation. By dissecting muscles in detail, he determined they were bundles of contractile fibres, not balloons that were inflated by an 'animating spirit'. The heart was not made of any special substance nor was it an internal cauldron of boiling blood but a muscle. As for the brain, René Descartes, the French mathematician and philosopher, had concluded the pineal gland sitting at its centre was the location of the soul and manipulated the body like a puppet. In a public lecture in 1665 in Paris, Steno demonstrated that this could not be the case, as the gland could not move and gyrate as Descartes had supposed.

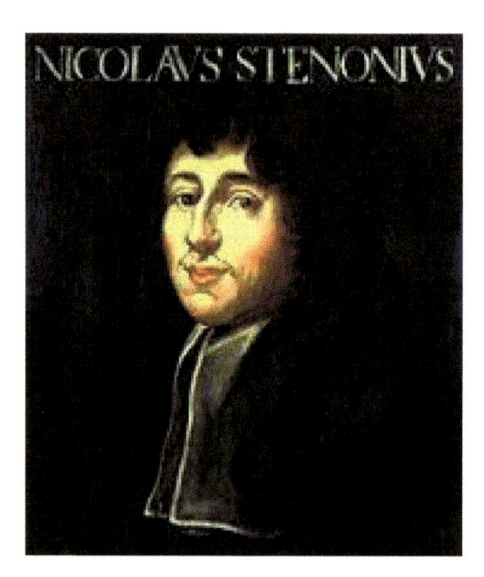

Portrait of Nicolas Steno (c. 1666–1667). Unsigned but attributed to Florence court painter Justus Sustermans.

These discoveries attracted the attention of Grand Duke Ferdinand de' Medici in Florence. Ferdinand had a strong interest in science, supporting Galileo and sponsoring an academy of his disciples. Steno was thus invited to his court as in-house physician, but also to educate and entertain.

So it came about that the shark caught by the Tuscan fisherman came to Steno in Florence. The dissection was routine, but Steno was struck by the similarity of the shark's teeth to certain medicinal stones called *glossopetrae* or tongue stones. These were thought to grow within rocks within the Earth, but Steno determined that the shark's teeth and the *glossopetrae* were very much the same — in other words *glossopetrae* were fossilized shark's teeth. This was reported in a paper published in 1667 (*Canis carchariae dissectum caput* – 'A Shark's Head Dissected') and spurred his interest in geology. With the support of Grand Duke Ferdinand, he proceeded to travel around Tuscany for the next few months to study its fossils and geology.

The result was his masterpiece published in 1669 — *De solido intra solidium naturaliter contento dissertationis prodomus* ('Preliminary discourse to a dissertation on solids naturally contained within a solid'). This set out some basic principles upon which the science of geology could be built. He recognized that rocks containing fossils had originally been soft sediments hardened into rock after burial. Therefore, fossils were organic in origin, rather than being odd facsimiles of living creatures and their constituent body parts that happened to be embedded in rock. This matched the observations of the English scientist Robert Hooke who had utilized the newly invented microscope to note that the microstructure of fossil wood was similar to charcoal (*Micrographia,* 1665).

The position of these fossils on land, often far from the sea and high above it, was less easy to explain. Steno favoured their deposition in sediments relating to the biblical flood, left behind when waters receded. In Steno's time, Earth history and human history were viewed

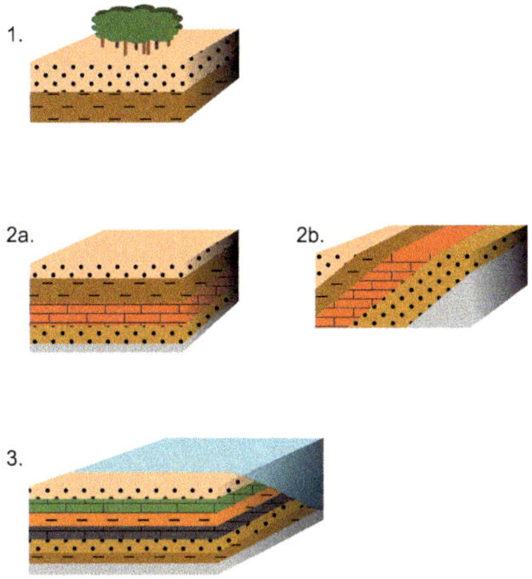

Key geological principals as set out by Steno in 1669.
1: Principal of Superposition - younger strata overlie older strata;
2: Principal of Original Horizontality – sedimentary layers are originally deposited flat (2a) and can therefore be tilted by crustal movements (2b);
3: Principal of Lateral Continuity – layers of sedimentary rock are continuous until they encounter other solid bodies that block their deposition or until they are acted upon by agents after deposition.

as effectively synchronous and the biblical flood seemed almost unbelievably ancient. To suggest that fossils could be so old was challenging — such were the perceptions of the age of the Earth at that time.

Having recognised how crystals and shells grow progressively by the addition of layers, in a leap of insight he applied this concept to the strata of the Earth. This is now known as the Principle of Superposition — in a sequence of sedimentary layers the bottom layer is the oldest and those above are progressively younger.

Steno also stated that water-deposited sediments are laid down horizontally and form continuous layers — the Principals of Original Horizontality and Lateral Continuity. Therefore, tilted or discontinuous strata represented folding or faulting movements within the crust, features he observed within Tuscany.

Steno had shown that the rock record and its fossil content could be used to interpret a chronology. That is to say, rocks represent a book waiting to be read that can reveal the history of our planet. This was a fundamental contribution to science. Such principles paved the way for James Hutton over 100 years later to appreciate the long age of the Earth and for William Smith, Georges Cuvier and others to begin to classify the strata of the Earth and initiate the unravelling of Earth's history.

In 1669, Steno was obliged to end his geological studies with a return to medical work in Copenhagen. This coincided with his conversion from Lutheranism to Catholicism and subsequently immersing himself in theology. He never returned to geology.

In 1675 he took holy orders and became a priest in 1677. He became a bishop tending to the small Catholic population in northern Germany. He took a vow of poverty and the rigors of this life destroyed his health, bringing about his death in 1686 at the relatively young age of 48. His legacy is the foundation of the sciences of paleontology and geology and critical insights into how the human body works. For his religious endeavors, he was beatified by Pope John Paul II in 1988.

REFERENCES

This essay has drawn upon information from the following sources:

Adams, F.D. 1938. *The Birth and Development of the Geological Sciences.* Williams & Wilkins.

Cutler, A. 2003. *The Seashell on the Mountaintop.* William Heinemann Ltd, 240pp.

Cutler, A. 2007. Nicolaus Steno: unlocking the Earth's geological past. In: Huxley, R. (ed.) *The Great Naturalists.* Thames & Hudson, 86-91.

Gohau, G. 1990. *A History of Geology.* Rutgers University Press. 259pp.

Gould, S.J. 1983. *Hen's Teeth and Horse's Toes.* W.W. Norton & Company, Inc., 413pp.

Oldroyd, D.R. 1996. *Thinking About the Earth.* The Athlone Press, 410pp.

Rudwick, M.J.S. 1972. *The Meaning of Fossils.* The University of Chicago Press, 287pp.

Rudwick, M.J.S. 2014. *Earth's Deep History.* The University of Chicago Press, 360pp.

Scherz, G. Undated. *Niels Steensen.* Royal Danish Ministry of Foreign Affairs. 95pp.

http://nielssteensen.dk/

Abraham Werner

Abraham Werner – an 1801 portrait by Christian Leberecht Vogel.

It would be easy to dismiss Abraham Werner from the inventory of Great Geologists because of his promotion of a controversial and severely flawed theory that became known as "Neptunism." The theory held that almost all rocks were the result of deposition or precipitation of sediments accumulating on the floor of an ancient ocean. Even during its late 18th century heyday, this notion was rejected by those who recognised the importance of igneous processes (volcanoes and magmatic intrusions). In his *Principals of Geology*, Sir Charles Lyell was particularly critical of Werner, although with some distortion of Werner's work. Nonetheless, any dismissal of Werner as a Great Geologist would ignore both his tireless promotion of geology as one of the most respected teachers in Europe and his development of the rudiments of mineralogy and stratigraphy. In short, 18th century geology as a science moved forward in no small part because of the observations and teachings of Werner.

Abraham Gottlob Werner was born in 1750, in the town of Wehrau, then located in Saxony (the town is now called Osiecznica and is within Poland). The local region has a long history of coal and metal ore mining. With family connections in these industries, he was sent to study law and mining at Freiberg Academy and Leipzig University. An apt scholar, on completion of his education in 1775, Werner became Inspector and Teacher of Mining and Mineralogy at the small, but influential, Freiberg Mining Academy. This would be his intellectual base for the next forty years.

Even before Werner's first professional appointment, he had published an important and influential textbook in 1774: *Von den äusserlichen Kennzeichen der Fossilien (On the External Characters of Fossils)*. Despite the title, the book was a systematic description of minerals, with a particular focus on their colours. Despite this promising start as an author, Werner produced few other publications. Instead, he preferred to concentrate on oral teaching and it was his students who published or otherwise promoted many of his ideas. Not only did Werner develop an aversion to writing, in later life, he adopted the practice of storing his mail unopened. Elected a foreign member of the French Académie des Sciences in 1812, he only learned of the honour much later, when he happened to read about it in a journal.

By all accounts, Werner was a brilliant teacher. He rapidly attracted bright minds from across Europe to attend his classes in Freiburg, including Robert Jameson, who later became Regius Professor of

Natural History at Edinburgh (and, in turn, taught Charles Darwin) and Alexander von Humboldt, one of the true greats of Natural History in the early 19th century. Jameson would become amongst the most vocal and articulate promoters of Werner's ideas.

Werner emphasised to his students the importance of making accurate geological observations in the field. According to Werner, rock units or "formations" (Gebirge: literally, mountains) had three dimensions (i.e. structural relationships to each other and to the topography, as well as a distinct stratigraphic order). In his geological thinking, Werner was much influenced by the publications of his fellow countryman, Johann Gottlob Lehmann. Lehmann had built on the Law of Superposition, as earlier described by Nicolas Steno, and emphasised the importance of layering within rock successions — what we now term "stratigraphy." Whilst Lehmann described stratigraphy on a local scale, Werner expanded these ideas to a global scale.

Werner proposed new ways of thinking about geologic formations, redefining formation to refer not just to the chemical and lithological makeup of a rock, but to the timing of its development. He defined formations as "bodies of rock laid down in the same period," giving scientists a new way of thinking about geological history. These ideas were first published in a booklet in 1787 — *Kurze Klassifikation und Beschreibung der verschiedenen Gebirgsarten (Brief Classification and Description of the Different Species of Formations)*.

By 1796, Werner believed that the stratigraphy of the Earth could be divided into five principal formations, mostly related to deposition within, or precipitation from, a primordial ocean:

1. Primitive (*Urgebirge*) Formation: granites, gneiss, schists and metasediments considered to be the first precipitates from the ocean

2. Transition (*Übergangsgebirge*) Formation: limestones, dykes, sills and thick sequences of greywackes with few or no fossils that were the first orderly deposits from the ocean

3. Secondary or Stratified (*Flötz Gebirge*) Formation: the remaining, obviously stratified and clearly fossiliferous, rocks. These were thought to represent the emergence of mountains from beneath the ocean and were formed from the resulting products of erosion deposited on their flanks.

4. Alluvial or Tertiary (*Aufgeschwemmte Gebirge*) Formation: poorly consolidated sands, gravels and clays formed by the withdrawal of the ocean from the continents

5. Volcanic (*Vulkanische Gebirge*) Formation: younger lava flows demonstrably associated with volcanic vents. Werner believed that these rocks reflected the local effects of coal combustion.

These five units were divided into a variety of sub-units, based on Werner's observations around Saxony. He is often criticised for developing a global stratigraphy from observations within only a small area of central Europe. However, it can be noted that he was an avid collector of geological literature and must have found support for his ideas in such readings.

The basic concept of Wernerian geology was the belief in an all-encompassing ocean that gradually receded to its present location, while precipitating or depositing almost all the rocks and minerals in the Earth's crust. The emphasis on this initially universal ocean spawned the term Neptunism, which became virtually synonymous with Wernerian teaching. A universal ocean led directly to the idea of universal formations, which Werner believed could be recognised on the basis of lithology and superposition. He used the term "geognosy" (meaning, knowledge of the Earth) to define a science based on the recognition of the order, position and relationship of the layers forming the Earth. As he commented towards the end of his life, *"Our Earth is a child of time and has been built up gradually."*

The notion that most rocks are the result of deposition in a primordial ocean is patently wrong. However, Werner's ideas can, nonetheless, be translated into aspects of the geological history of the Earth and its associated stratigraphy that we know today. His Primitive and Transitional Formations may represent Precambrian basement and sediments, respectively, whilst his fossiliferous Secondary (Flötz) Formation may represent much of Phanerozoic sedimentation. His Alluvial and Volcanic formations represent relatively recent rocks.

Although erroneous in his understanding of the origin of many rocks, Werner's global stratigraphy paved the way for the more detailed stratigraphic thinking of the 19th century.

A principal focus of Neptunism that provoked almost immediate controversy involved the origin of basalt. Basalts, particularly those occurring as horizontal sills, were differentiated from surface lava flows, and the two were not recognised as the same rock type by Werner. Lavas and volcanoes of obviously igneous origin were treated as very recent phenomena, unrelated to the universal ocean that formed the majority of layers of the Earth. Werner believed that volcanoes only occurred in proximity to coal beds and that lavas were the result of the combustion of coal. Basalt at high elevations, interbedded with sedimentary rocks, such as sandstones and shales, proved to Werner that they were chemical precipitates of the ocean. Vertical dykes were considered to be precipitates from the universal ocean infilling fissures.

These views were countered by, amongst others, the Scottish geologist James Hutton. Hutton favoured the notion that granites and basalts were the products of heat within the Earth creating molten magma (Plutonism). His observation of cross-cutting intrusions demonstrated this and was a major factor in the discrediting of Neptunism, along with observations of modern and ancient volcanoes documented by one of Werner's former students, Leopold von Buch.

A second controversy surrounding Neptunism involved the volumetric problems associated with the existence of a universal ocean. How could Werner account for the covering of the entire Earth, and then the shrinking of the ocean volume, as the primitive and transition mountains emerged and the secondary and tertiary deposits were formed? The movement of a significant volume of water into the Earth's interior had been proposed by the classical Greek geographer, Strabo, but this was not embraced by Werner because it was associated with conjecture. Nevertheless,

The location of Scheibenberg, as photographed c. 1900. This was a key outcrop in the development of Werner's theory of the Earth. The vertically jointed basalts visible on the hill are interbedded with sedimentary rocks and, in the view of Werner, have gradational contacts. This demonstrated the precipitation of basalt from a universal ocean, according to Werner.

with his views on basalt, Werner obviously did not believe that the interior of the Earth was molten. He considered that the primeval, original ocean might have been drawn away by the attraction of a celestial body passing near the Earth (obviously, a conjecture that he was happy to embrace). However, he did not emphasise this point and never really explained how the original ocean had shrunk to the size it is today.

Werner was plagued by frail health his entire life, and led a quiet existence in the immediate environs of Freiberg. An avid mineral collector in his youth, he abandoned field work altogether in his later years. There is no evidence that he had ever travelled beyond Saxony in his adulthood. He died at Dresden in 1817 from stress-related complications said to have been caused by his consternation over the misfortunes that had befallen Saxony during the Napoleonic Wars. Although primarily a mineralogist and mining geologist, Werner was at the forefront of promoting geology as a history of the Earth and, in so doing, illuminated the way forward for the great advances of the science in the 19th century.

REFERENCES

This essay has drawn upon the following works:

Adams, F.D. 1938. *The Birth and Development of the Geological Sciences*.

Greene, M.T. 1982. *Geology in the Nineteenth Century*. Cornell University Press. 324pp.

Hallam, A. 1983. *Great Geological Controversies*. Oxford University Press. 244pp.

Laudan, R. 1987. *From Mineralogy to Geology*. The University of Chicago Press. 278pp.

Oldroyd, D.R. 1996. *Thinking About the Earth*. The Athlone Press, 410pp.

Rudwick, M.J. 1997. *Georges Cuvier, Fossil Bones, and Geological Catastrophes*. The University of Chicago Press. 301pp.

Rudwick, M.J.S. 2005. *Bursting the Limits of Time*. The University of Chicago Press, 708pp.

Freiburg Institute of Mineralogy.

James Hutton

One day in 1788 three eminent gentlemen intellectuals from Edinburgh set off in a boat along the coast of south-eastern Scotland, south of Dunbar and not too far from the English border. They landed at Siccar Point, a rocky promontory jutting into the North Sea. One of the party expounded to the others his interpretation of the rock formations forming the foreshore in front of them. He observed near-vertically arranged grey shale beds overlain by reddish sandstones lying at an angle closer to horizontal, thus the shales and sandstones were separated from each other by a distinct angular discordance. To the speaker in the landing party, this succession implied that the shales had been deposited on an ancient sea-bed over a long period of time, that they had then been buried and solidified, then uplifted by forces within the Earth, tilted and eroded, and then the sandstones had been deposited upon the eroded surface and then in turn solidified and uplifted. This process must have taken a huge amount of time, much longer than the commonly accepted wisdom in the 18th century that the Earth was a few thousand years old based on a literal interpretation of biblical events. John Playfair, the celebrated mathematician who was at Siccar Point that day, wrote that on hearing of this explanation that the *"the mind seemed to grow giddy by looking so far into the abyss of time"*. The person who gave the explanation that this rock succession implied the vastness of geological time was James Hutton – the founder of modern geology.

James Hutton was born in Edinburgh in 1726 and although he studied medicine, he developed interests in chemistry, meteorology and agriculture. He displayed an intellectual curiosity into a diversity of subjects that typified many of the Scottish Enlightenment scholars in the second half of the 18th century. He developed a means to extract salts from coal soot that could be used in cloth dyeing (and earned a healthy income from this), wrote a *"Theory of Rain"* that preceded much modern meteorological thought on the hydrological cycle, and acquired a large farming estate which encouraged him to study modern agricultural methods. It was farming that really ignited his interest in geology as he noted the formation of soils and their erosion and subsequent transport into the sea by streams was a progressive, long-term, process and one that was cyclic – the sediments that were deposited were formed into rocks, uplifted and eroded again.

By 1753 he was able to write *"I have become very fond of studying the surface of the Earth and am looking with anxious curiosity into every pit or ditch or river bed that falls my way"*. He was now travelling far and wide in England, Scotland and Wales to make geological observations and as he did so he collected fossils and observed what we would call today sedimentary structures (such as ripple marks) giving him the idea that some rocks were the product of deposition in ancient seas and that sedimentary processes that were taking place on Earth today had taken so in the past. This was the concept of Uniformitarianism or "the present is the key to the past" that was later expounded upon by Sir Charles Lyell. Moreover, Hutton viewed the history of the Earth to be markedly cyclic – it could not simply be progressively eroding away since its creation. Instead there were repeated cycles of erosion,

James Hutton painted by Sir Henry Raeburn in 1776.

The foreshore at Siccar Point visited by Hutton in 1788. Vertically bedded grey Silurian deep-marine turbidites are separated by a prominent unconformity from the overlying Devonian fluvial red sandstones. "Hutton's Unconformity" is a place of pilgrimage for many geologists, especially those interested in stratigraphy.

deposition and uplift – the creation of a succession of habitable worlds in his view. To explain the phases of uplift Hutton observed igneous intrusions into sedimentary rocks suggesting to him the remarkable (for the time) idea that the interior of the Earth was hot, that molten intrusive rocks were the product of this process and that the expansive power of heat from within the Earth could be responsible for the uplift – a forerunner of today's tectonic theories.

Hutton spent over 30 years gathering his ideas and in 1785 first read his *"Theory of the Earth"* to the Edinburgh Royal Society. We can only imagine the reaction at the time to the radical thoughts on the history of the Earth and processes on and within it that Hutton outlined. He published his theories in 1788 and continued to gather evidence to support them (the boat trip to Siccar Point was one such episode). By 1795 he was able to publish a much longer thesis (including a description of natural selection that preceded Darwin by 50 years), but sadly the articulation of his ideas was not always the clearest in a book that ran to over 2100 pages. After his death in 1797, his friend John Playfair published in 1802 a much more accessible *"Illustration of the Huttonian Theory of the Earth"* and the progression of the science of geology was well underway.

James Hutton in the field in an 18th century sketch.

REFERENCES

This essay has drawn upon information from the following sources:

Adams, F.D. 1938. *The Birth and Development of the Geological Sciences*. Williams & Wilkins.

Baxter, S. 2003. *Revolutions in the Earth*. Weidenfeld & Nicolson.

Cook, J. 2007. James Hutton: discoverer of geological time. In: Huxley, R. (ed.) *The Great Naturalists*. Thames & Hudson, 186-189.

Gohau, G. 1990. *A History of Geology*. Rutgers University Press. 259pp.

Gould, S.J. 1983. *Hen's Teeth and Horse's Toes*. W.W. Norton & Company, Inc., 413pp.

Gould, S.J. 1987. *Time's Arrow, Times's Cycle*. Harvard University Press, 222p.

Greene, M.T. 1982. *Geology in the Nineteenth Century*. Cornell University Press. 324pp.

Hallam, A. 1983. *Great Geological Controversies*. Oxford University Press, 244pp.

Laudan, R. 1987. *From Mineralogy to Geology*. The University of Chicago Press. 278pp.

McIntyre, D.B. & McKirdy, A. 1997. *James Hutton: The Founder of Modern Geology*. The Stationery Office.

McPhee, J. 1998. *Annals of the Former World*. Farrar, Straus and Giroux. 696pp.

Oldroyd, D.R. & Hamilton, B.M. 2002. Themes in the early history of Scottish geology. In: Trewin, N.H. (ed.) *The Geology of Scotland*. The Geological Society, London, 27-44.

Oldroyd, D.R. 1996. *Thinking About the Earth*. The Athlone Press, 410pp.

Repcheck, J. 2003. *The Man who Found Time*. Simon and Schuster.

Rudwick, M.J.S. 2005. *Bursting the Limits of Time*. The University of Chicago Press, 708pp.

Rudwick, M.J.S. 2014. *Earth's Deep History*. The University of Chicago Press, 360pp.

http://blogs.plymouth.ac.uk/sustainableearth/towards-a-sustainable-earth/

Pinkish granitic intrusions into grey Dalradian metasediment at Glen Tilt in Perthshire, Scotland. Hutton's observation of these phenomena in 1785 supported his ideas that the interior of the Earth was hot and that igneous rocks were the product of molten magmas associated with this heat and could, under circumstances of immense pressure, intrude into older rocks.

William Smith

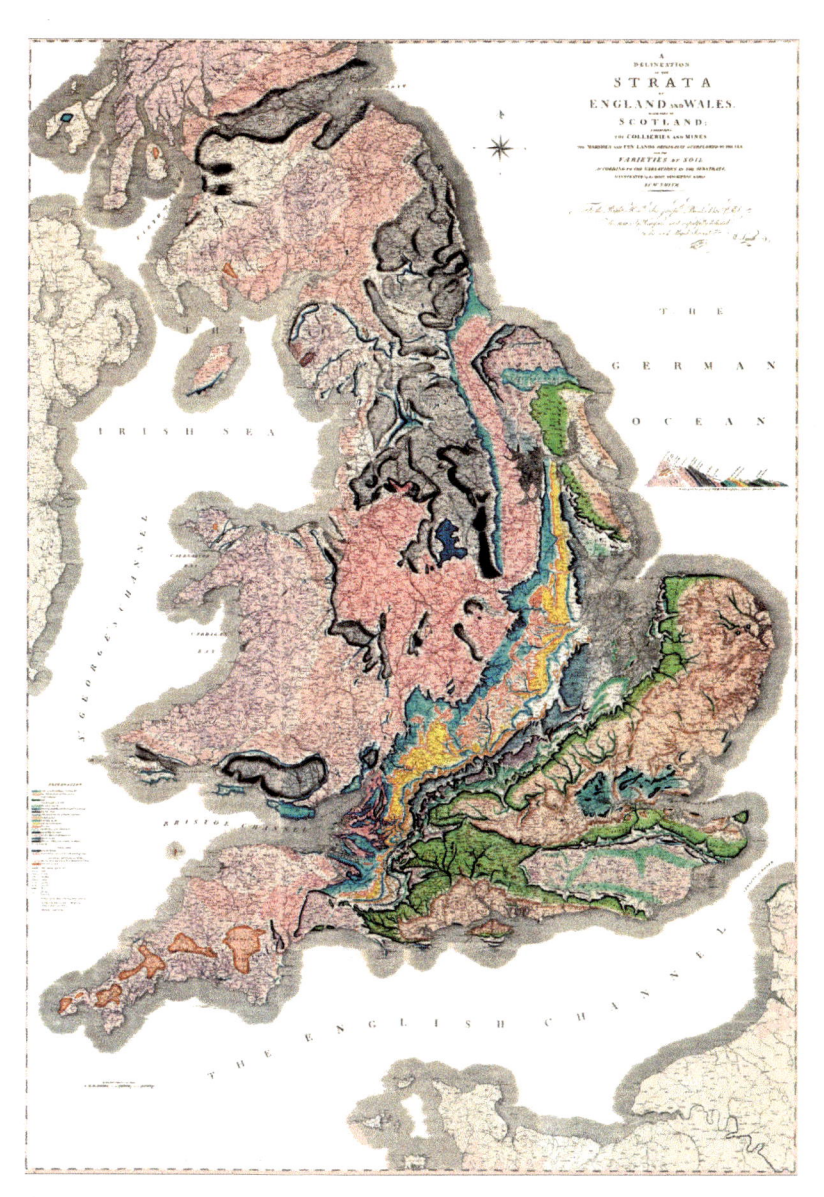

Smith's 1815 geological map of England, Wales and part of Scotland.

Practical geologists make maps. Not just those showing the rocks that occur at surface in a region, but also depth maps, thickness maps, paleogeographic maps, amplitude extraction maps - we make all of these and many more besides with pencils and paper or ever-increasingly in 2-D and 3-D software packages. The first major effort at creating a geological map of a country was undertaken by a practical geologist - William Smith - who was engaged in surveying canal routes in the late 18th and early 19th centuries. His masterpiece was the first geological map of England and Wales published just over 200 years ago. To create the map Smith needed to identify and correlate strata and found that the fossil content of sedimentary rocks was an ideal tool for doing so. He was therefore a key pioneer in biostratigraphy or practical paleontology.

Smith was born in 1769 in the beautiful Oxfordshire village of Churchill. In sight of the Cotswold Hills and surrounded by fossiliferous Jurassic rocks it is no surprise that he became interested in geology from an early age. In contrast to many notable scientists of the late 18th and early 19th centuries, Smith came neither from a privileged background (he was the son of the village blacksmith) nor was he a particularly adept academic scholar. Instead, by the age of 18, he was engaged in learning to be a surveyor – a practical discipline in which he excelled. Economic developments in the late 18th century provided ample opportunities for a talented surveyor. The Enclosure Acts set about organising the English countryside into the ordered network of fields and hedgerows we know today and at the same time the development of steam-driven machinery provided a ready need for coal and for canals to transport it and other goods. Smith was a busy man and by 1791 he had moved to the county of Somerset and was engaged in a study of a working colliery to delimit its extent and to maximise the extraction from the Carboniferous Coal Measures a few hundred feet below the surface. It was here that Smith's understanding of stratigraphy began to take hold. He noted a distinct order to what we would today term as cyclothems or parasequences of coal, mudstone and sandstone and that individual rock units could be distinguished by their lithology and fossil content. This he recorded in outline in a brief note now preserved in the Oxford University Geological Museum entitled "*Original Sketch and Observations of My First Subterranean Survey of Mearns Colliery in the Parish of High Littleton*". It can be argued that this is the founding document of practical stratigraphy, but much more was to come from Smith.

From 1793–1799 he was chiefly engaged in surveying the route for the Somerset Coal Canal. This cut through what we now recognise as Carboniferous, Triassic and Jurassic sediments. Smith had the skill to recognise the distinct order of this stratigraphy as it dipped gently eastwards and that certain beds could be characterised by certain fossils. By 1799, encouraged by his society friends in the well-heeled city of Bath, he presented his first geological map of the area surrounding the city, and also his table of strata in the region – "*Order of the STRATA and their Embedded ORGANIC REMAINS, in the vicinity of BATH; examined and proved prior to 1799*". Note that the title of this historic document provides a link between paleontology and stratigraphy – that rocks could be characterised and correlated by their fossil content.

Smith was now travelling widely in England and Wales and expanded his observations from Somerset to the whole of these countries and southern Scotland. By 1801 an unpublished outline geological map of England was prepared, but it took Smith until August 1st 1815 to see the publication of 400 beautifully hand-coloured copies of his map *"A Delineation of The Strata of England and Wales with part of Scotland"*. A cross-section accompanied the map. There are at least two things remarkable about this map which can be viewed in the entrance hall of the Geological Society of London. Firstly that it is the work of just one man. He had no geological survey at his disposal. It is a record of his observations alone. Secondly, is that it is remarkably accurate given that he was working alone and had deduced for himself the connection between fossil content and stratigraphy - "putting paleontology on the map" - so to speak. In the course of his work he collected thousands of fossil specimens. Subsequent to the map, he published plates of the key fossils from the main stratigraphic units he recognised.

Portrait of William Smith by Hugues Fourau, 1837.

It has been well-recorded that fate was not kind to this pioneer of geology. He was constantly troubled by financial difficulties and in 1819 he spent a short period of time in a debtor's prison. Moreover it took over 15 years after publication of the 1815 map for the embryonic geological establishment to recognise his contribution to the science. No doubt this was in part because of his social background (despite his attempts, which led to debt, to "rise above his station") but also because he was not a man for theory. He made his map and utilised what we would now term as biostratigraphy because it was a practical means of helping him with his surveying work. He did not speculate on why strata are ordered in a particular way or why fossil content changed between particular rock units. He was effectively doing what many geologists do in practice to this day, making empirical observations and relying on specialists or new data for explanations. Geology is a practical subject and Smith has been followed by a proud line of industry-based geologists who have made their contribution to the science. There is a famous quote by the great British geologist H.H. Reid: *"The best geologist is the one who has seen the most rocks"*. Smith spent almost his whole life gathering geological data and putting it to practical use. Moreover, by doing do he greatly advanced geology as a science.

REFERENCES

This essay has drawn upon information from the following sources:

Adams, F.D. 1938. *The Birth and Development of the Geological Sciences*. Williams & Wilkins.

Morton, J.L. 2007. William Smith: the father of English geology. In: Huxley, R. (ed.) *The Great Naturalists*. Thames & Hudson, 218-223.

Oldroyd, D.R. 1996. *Thinking About the Earth*. The Athlone Press, 410pp.

Rudwick, M.J.S. 1972. *The Meaning of Fossils*. The University of Chicago Press, 287pp.

Rudwick, M.J.S. 2005. *Bursting the Limits of Time*. The University of Chicago Press, 708pp.

Torrens, H.S. 2001. Timeless order: William Smith (1769-1839) and the search for raw materials 1800-1820. In: Lewis, C.L.E. & Knell, S.J. (eds.) *The Age of the Earth: from 4004 BC to AD 2002*. Geological Society, London, Special Publications, 190, 61-84.

Winchester, S. 2001. *The Map that Changed the World*. Viking.

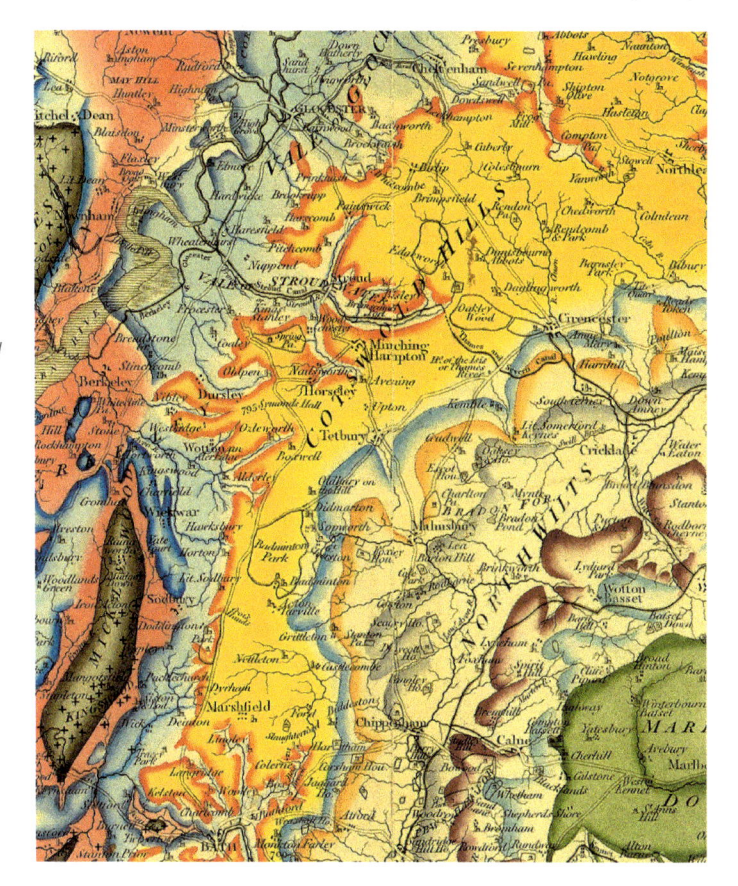

Detail of Smith's 1815 geological map of England, Wales and southern Scotland.

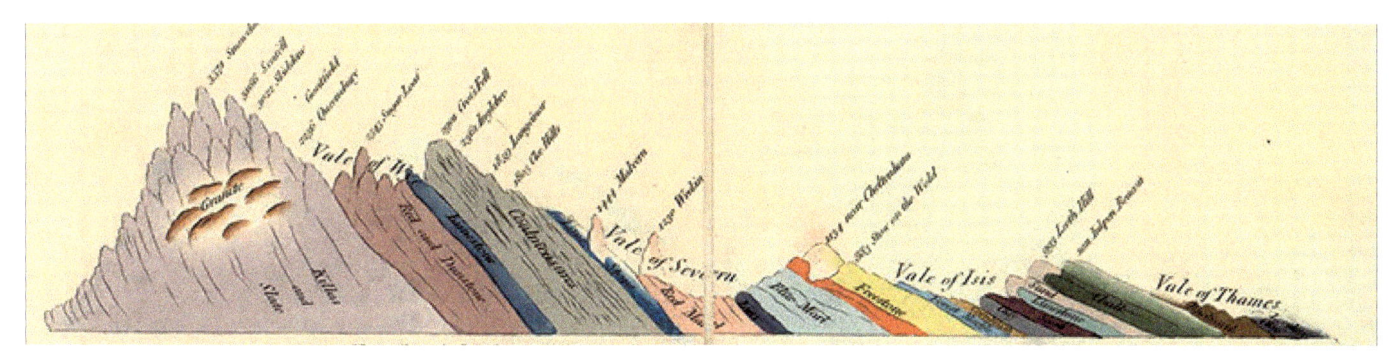

Cross-section across southern England that accompanies Smith's 1815 map.

Georges Cuvier. Painted by W.H. Pickersgill, 1831. Engraved by George T. Doo, 1840.

Baron Georges Cuvier

One of the finest minds of the Age of Enlightenment was that of Georges Cuvier, yet he is often portrayed negatively in histories of geoscience because of his opposition to evolution and his promotion of a history of the Earth often described as Catastrophism. Notwithstanding the rights and wrongs of such judgments, they belittle the massive contributions he made to the emerging science of geology in the early 19th century. These included: the recognition that the vast majority of fossils are the remains of extinct creatures; the use of comparative anatomy to resolve the form of complete creatures from their fragmentary fossil remains; the exposition of the value of fossils in stratigraphic classification and correlation; and pioneering attempts to interpret stratigraphic successions as a history of changing paleoenvironments.

In 1812 he wrote *"We admire the power by which the human mind has measured the movements of the globes, which nature seemed to have concealed forever from our view; genius and science have burst the limits of space, the observations interpreted by reason have unveiled the mechanism of the world. Would there not also be some glory for man to know how to burst the limits of time and, by observations, to recover the history of this world and the succession of events that preceded mankind's birth?"* This was a rallying call for geology to match the achievements of physics and astronomy, but in particular, to the purpose of stratigraphy and Cuvier was amongst the first to imagine a succession of past worlds, populated by mostly extinct organisms. These achievements stand alongside the colossal work he carried out in the comparative anatomy and classification of living creatures, which mark him out as one of the greats of biology and of science in general.

The son of a military man, Cuvier was born in 1769 in the town of Montbéliard, now in eastern France. He studied in Stuttgart, where he became interested in entomology, botany and zoological classification. In 1788 he moved to Caen in Normandy, where he was employed as a teacher to an aristocratic family. The work was not arduous, so he expanded his scientific interests to the description of marine animals, molluscs and arthropods. His location in rural Normandy was, perhaps, fortunate as he was not drawn into the French Revolution of 1789 and continued his work as a naturalist of increasing reputation, such that by 1795 he was invited to Paris and became a teacher of natural history at one of the new Écoles Centrales.

He progressed rapidly through French scientific society. At the Muséum d'Histoire Naturelle, he created a stunning gallery devoted to the animal world. With over 16,000 zoological specimens organised by zoological class and illustrating the relationship between form and function, this gallery became one of the scientific sensations of the early 19th century.

Having examined, dissected and drawn almost every known living animal, in 1817 he published an inventory of the animal kingdom and a classification based on the functional morphology of each creatures' bones and organs. This four volume work was entitled *Règne Animal Distribué d'après son Organisation* — a second edition was published in 1829-30 in five volumes. This is undoubtedly one of the masterpieces of zoological science.

His study of the comparative anatomy of living creatures naturally brought his attention to fossils, initially extinct vertebrates. In 1796, he recognised that an engraving of a huge fossil skeleton from South America, housed in the royal museum in Madrid, was an extinct giant sloth, which he named *Megatherium* ("huge beast"). Soon afterwards he demonstrated that fossil mammoth bones and teeth from Siberia, although similar to elephants from Asia and Africa, belonged to a separate, extinct species. He determined that the spectacular skull of a huge fossil animal from a chalk quarry in Maastricht had belonged to a large marine lizard (now called *Mosasaurus*) and a small fossil reptile skeleton from Jurassic strata in Bavaria was that of a flying reptile, which he called *Ptero-dactyle*. Such determinations were made possible by Cuvier's detailed understanding of the anatomical function of the bones preserved, from which he inferred the nature of a complete creature and its mode of life. Most importantly, Cuvier recognised the existence of creatures that were not living today, which for him gave evidence of worlds previous to todays, inhabited by creatures very different to those around us. Extinction was a real feature of the natural world, so to explain this required consideration of the geological past.

Accordingly, Cuvier's geological interests rapidly expanded beyond the determination of the nature of fossils. In collaboration with the mineralogist Alexandre Brongniart, in 1811 he presented *Essai sur la géographie minéralogique des environs de Paris*, which contained not only a geological map of the Paris Basin, but a stratigraphic synthesis in the form of a novel sedimentary log. In the manner of William Smith in Britain, they used the fossil content of the strata for subdivision and correlation. Moreover, they determined a succession of alternating freshwater and marine environments — a pioneering attempt to reconstruct the geological history of the succession. The two men progressively refined their stratigraphic understanding with more detailed logs, maps and descriptions, the last appearing in 1832 at the time of Cuvier's death.

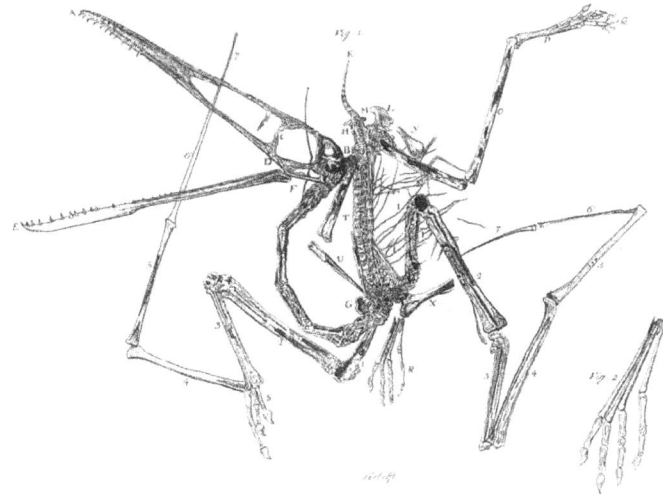

The fossil specimen that Cuvier recognised as a flying reptile - *Ptero-dactyle*.

This study of geological history, punctuated by rapid environmental change, led Cuvier to suggest that Earth's history had been episodically interrupted by sudden 'revolutions' that caused the extinction of existing fauna. Such ideas were first documented in *Discours Préliminaire* of his great work of 1812, *Recherches sur les Ossemens Fossiles*, which was later published separately as *Discours sur les Révolutions de la surface du Globe* and which became known as the definitive catastrophist view of geological history. Ultimately these views came into conflict with Charles Lyell's uniformiatarism, a steady-state view of Earth history.

The impact of *Discours* upon early 19th century science was profound. Here was an attempt to elucidate a history of the Earth that went beyond the observations and correlations of his British contemporaries such as William Smith. While Smith may have expressed the power of fossils for stratigraphic subdivision, correlation and mapping slightly ahead of Cuvier and Brongniart, it was Cuvier who first attempted to determine the history of our planet through the vastness of geological time. Cuvier's views can be summarised by this translated paragraph:

"Life upon the earth in those times was often overtaken by these frightful occurrences. Living things without number were swept out of existence by catastrophes. Those inhabiting the dry lands were engulfed by deluges, others whose home was in the waters perished when the sea bottom suddenly became dry land; whole races were extinguished leaving mere traces of their existence, which are now difficult to recognise, even by the naturalist. The evidences of those great and terrible events are everywhere to be clearly seen by anyone who knows how to read the record of the rocks."

It is hard not to draw an analogy between Cuvier's view of geological history and the revolutionary nature of contemporary politics in Europe, especially France where, at the end of the

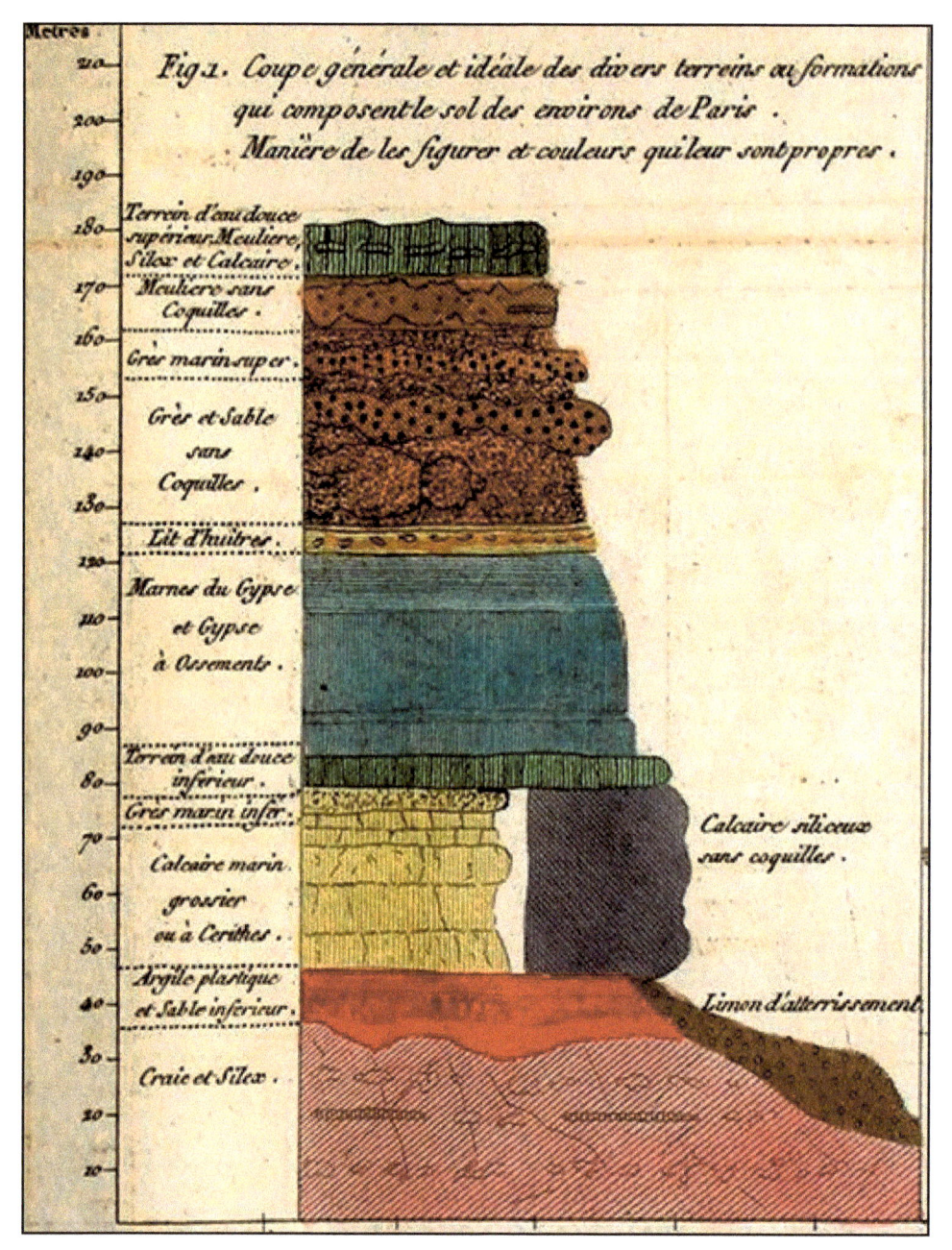

The "ideal section" of the geological succession of the Paris Basin as summarized by Cuvier and Brongniart in 1811. This novel geological log was the basis for reconstructing the succession as a series of changing paleoenvironments marked by abrupt boundaries. This in turn contributed to Cuvier's ideas that the Earth history was punctuated by revolutions or catastrophes. Modern sequence stratigraphers will not be surprised that the boundaries between environmental change are abrupt given the "up systems tract" position of the succession.

18th century, the old order was being swept away by the new. Regardless of whether Cuvier was influenced by political events around him, his theory was born of scientific observations. If, for example, his species of fossil mammoth had been perfectly adapted to the cold conditions it lived in, why did it become extinct? Cuvier felt sure that only a major environmental catastrophe could be responsible.

Given that Cuvier perceived Earth history to be a series of lost worlds each with their own distinctive fauna, it is not surprising that he was opposed to the early theories of evolution (for example, Jean-Baptiste Lamarck's 1809 work, *Philosophie Zoologique*) expressed as the transmutation of species in a state of flux. In Cuvier's view the fossil record lacked evidence for transitional types. Moreover, he viewed each species as a perfect anatomical machine, suited to its life and environment. Minor modifications would most likely destroy this perfection and lead to extinction.

Cuvier's description of Earth history as a series of former worlds inhabited by exotic extinct creatures then destroyed by catastrophes (the last being the biblical flood) had a profound impact on not only early 19th century science, but the public that heard him lecture in Paris or read his more accessible work (which was translated into several languages). This generated a popular, romantic interest in geology and paleontology, which was otherwise regarded as a very practical subject. Here was a scientific challenge for the age — to reconstruct and place in order these lost worlds — in other words, to determine the history of our planet back into deep time.

The challenge for the geologist to reconstruct the geological history of a region and to recognise the environments, and the fauna and flora that inhabited them, was set in motion by Cuvier. His views on catastrophes in the geological past are back in vogue — both as the mass extinction events that punctuate the geological record and as the relatively abrupt changes in relative sea-level that form the basis for modern sequence stratigraphy. His ideas are still current today.

In addition to his prolific scientific contributions, Cuvier was also an able administrator and adviser to governments. He served as an imperial councillor under Napoleon, president of the Council of Public Instruction, Grand Officer of the Legion of Honour, Minister of the Interior, and president of the Council of State under Louis Philippe. In 1819, he was created a peer for life in honor of his scientific contributions. He died in Paris in 1832 during an epidemic of cholera, forever to be remembered as one of France's greatest scientists.

REFERENCES

This essay has drawn upon information from the following sources:

Adams, F.D. 1938. *The Birth and Development of the Geological Sciences*. Williams & Wilkins.

Ager, D. 1993. *The New Catastrophism*. Cambridge University Press, 231pp.

Cadbury, D. 2000. *The Dinosaur Hunters*. Fourth Estate Ltd, 374pp.

Gohau, G. 1990. *A History of Geology*. Rutgers University Press. 259pp.

Gould, S.J. 1983. *Hen's Teeth and Horse's Toes*. W.W. Norton & Company, Inc., 413pp.

Greene, M.T. 1982. *Geology in the Nineteenth Century*. Cornell University Press. 324pp.

Hallam, A. 1983. *Great Geological Controversies*. Oxford University Press, 244pp.

Laudan, R. 1987. *From Mineralogy to Geology*. The University of Chicago Press. 278pp.

Oldroyd, D.R. 1996. *Thinking About the Earth*. The Athlone Press, 410pp.

Rudwick, M.J.S. 1972. *The Meaning of Fossils*. The University of Chicago Press, 287pp.

Rudwick, M.J.S. 1997. *Georges Cuvier, Fossil Bones, and Geological Catastrophes*. The University of Chicago Press. 301pp.

Rudwick, M.J.S. 2005. *Bursting the Limits of Time*. The University of Chicago Press, 708pp.

Rudwick, M.J.S. 2014. *Earth's Deep History*. The University of Chicago Press, 360pp.

Taquet, P. & Paidan, K. 2004. The earliest known restoration of a pterosaur and the phiolosophical origins of Cuvier's Ossemens Fossiles. *Compte Rendus Paleovol*, 3, 157-175.

Taquet, P. 2007. Georges Cuvier: extinction and the animal kingdon. In: Huxley, R. (ed.) *The Great Naturalists*. Thames & Hudson, 202-211.

Monmouth Beach at Lyme Regis.

Mary Anning

One of the pleasures of a trip to London is an opportunity to visit the Natural History Museum, one of the world's great treasure houses of zoology, mineralogy and paleontology. Amongst its many wonders is a corridor called "Marine Fossil Reptiles", along the walls of which are numerous specimens, remarkable for the beauty of their preservation, of skeletons of ichthyosaurs, plesiosaurs and other associated fossils from the Early Jurassic of southern England. Some of the most spectacular specimens were collected by a remarkable woman and pioneering fossil collector in the early 19th century – Mary Anning.

Mary was born in 1799 in the town of Lyme Regis on what is now celebrated as the Jurassic Coast of southern England. Lyme Regis attracts many visitors today, both professional and amateur, interested in the fossils that are abundant in the Early Jurassic sediments forming the cliffs and shoreface either side of the town. Indeed Lyme Regis was a centre for those interested in "curiosities" when Mary was born – her cabinet-maker father ran a side-line in selling "snake stones"(ammonites), "ladies fingers" (belemnites) and the like to well-heeled visitors. Unfortunately, her father died when she was only ten and the family struggled to make ends meet. Thus, necessity drove her to develop with her mother and elder brother Joseph the family business in collecting and selling fossils.

The romanticised notion of Mary Anning is that when only a young girl of about the age of eleven or twelve she found the first-ever ichthyosaur skeleton and sold it to save the family from destitution. In fact there was much more to Mary's geological contribution than this slightly exaggerated story. It was clear that she had an exceptional eye for locating and excavating spectacular fossil specimens and together with her brother she did indeed excavate a well-preserved ichthyosaur skeleton when she was young (and sold it for £23, about £600 in today's money – it's interesting to note that such skeletons would sell for many thousands of pounds today). She made four other major discoveries – in 1823 she found the first complete plesiosaur which was followed by a further specimen in 1830 that was sold for 200 Guineas. Both, along with famous ichthyosaur skull from her youth are now in the Natural History Museum. In 1828 she made the first discovery of a pterosaur in Britain – this "flying dragon" excited the public imagination and secured her fame as an exceptional fossil collector. In 1829 she discovered a new type of fossil fish, later named *Squaloraja*. Alongside these major contributions to the understanding of Jurassic life, she found further ichthyosaurs and a variety of fossil fish, crustaceans and molluscs. Because the collectors of important fossils in the early 19th century were rarely credited for their discoveries in the subsequent scientific description, there are probably many fossils found by Mary Anning that exist in museums, but we cannot be sure can be attributed to her. This is evident from this published remark on Mary by Bristol-based fossil enthusiast George Cumberland in 1823: *"This persevering female has for years gone daily in search of fossil remains of importance at every tide, for many miles under the hanging cliffs at Lyme….to her exertions we owe nearly all the fine specimens of Ichthyosauri of the great collections…"*

The early 19th century saw the first blossoming of the science of geology. With the discovery and promotion of deep time by James Hutton and others, the significance of fossils as more than trivial "curiosities" became clear. They helped understand Earth history and gave an insight into past life on the planet. So the fossils that Mary Anning found in the cliffs around Lyme Regis were of great

Depiction of Mary Anning finding her first fossil icthyosaur whilst still a young girl.

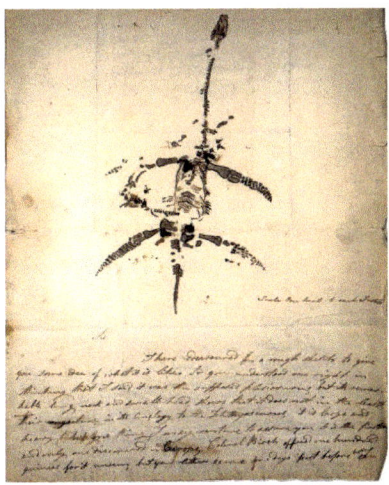

Letter and drawing from Mary Anning announcing the discovery of the first complete Plesiosaurus, 26 December 1823.

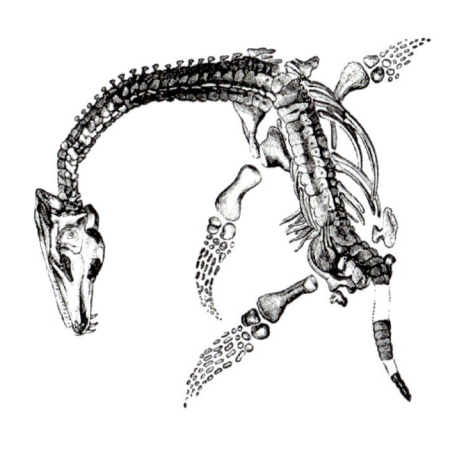

Drawing by William Buckland of the Plesiosaur discovered by Mary Anning in 1830.

interest to the gentlemen geoscientists of the time. They visited her and she regularly corresponded with many of them, Mary showing an aptitude for understanding the significance of her finds. Because of her major finds of spectacular fossils and her interaction with the gentry she enjoyed something of a celebrity status (even the King of Saxony visited her), despite her poor upbringing and being a woman in what was very much a man's world.

History has painted varying pictures of Mary's contribution to geology. Some historians of science have condemned her with faint praise as being simply a talented amateur. In fact she was a professional – her livelihood depended on her ability to locate, excavate, reassemble and preserve the fossils she found. And by corresponding with other geologists of the day she contributed greatly to the embryonic sciences of paleontology, paleoecology and stratigraphy.

True scientific recognition was denied her in her lifetime. Perhaps her much-remarked upon pride and forthright and business-like manner did not help. Her friend Anna Maria Pinkney wrote that Mary felt that *"the world has used her ill and she does not care for it, according to her account these men of learning have sucked her brains, and made a great deal by publishing works, while she derived none of the advantages…"*

Nonetheless, in the later years of her life pensions provided for by the British Association for the Advancement of Science and the Geological Society of London made sure she was not in need of income after her fossil collecting days were over.

She died in 1847 of cancer and after her death Henry De la Beche, the President of the Geological Society of London, paid great tribute to her in his presidential address. The value of her discoveries came to be truly appreciated in subsequent years and she progressively became regarded as an icon not just of geology, but of pioneering female contributions to science. Books about her life have multiplied. In fact should you visit the Natural History Museum during the school holidays you might just meet her talking about her fossil discoveries of two hundred years ago. Or at least an actress with a remarkable likeness who conveys the excitement of paleontology and geology to our younger generation.

REFERENCES

This essay has drawn upon information from the following sources:

Cadbury, D. 2000. *The Dinosaur Hunters*. Fourth Estate Ltd, 374pp.

Emling, S. 2009. *The Fossil Hunter*. St Martin's Griffin, New York. 234pp.

Pierce, P. 2006. *Jurassic Mary*. Sutton Publishing. 238pp.

Tickell, C. 2007. Mary Anning: fossil hunter. In: Huxley, R. (ed.) *The Great Naturalists*. Thames & Hudson, 250-254.

Torrens, H. 1995. Mary Anning of Lyme; the greatest fossilist the World ever knew. *British Journal of the History of Science*, 28, 257-284

William Buckland

Oxford University has a rich heritage of geological research and teaching. The first Reader of Geology at the university was the charismatic, some might say eccentric, William Buckland. His appointment at the university made him the first professionally appointed British geologist. Buckland's contributions to geology far transcend those of his university appointment. He collected and described a great number of fossils including the jaw of what would subsequently become to be understood as a dinosaur. He pioneered the understanding of cave faunas and discovered the remains of what would be one of the oldest humans in Europe. In his later career he was much taken with the importance of glaciation in shaping the landscape. Possessing a degree in divinity he spent much of his career attempting to reconcile the geological and biblical records. By doing so he popularized geology and paved the way for the flowering of the science in the mid-19th century.

Buckland was born in Axminster, Devon in 1784. In 1801 he won a scholarship to Corpus Christi College in Oxford and studied divinity with the intention of becoming an Anglican clergyman. Having obtained his degree he became a fellow of the college in 1809 and was ordained in the same year. Notwithstanding his theological interests an obsession with geology was taking root. He made numerous field excursions around the British Isles collecting fossils and noting the rock types present. Such interests led the university to make him Reader in Minerology in 1813. By 1819, with no less a personage than the Prince Regent as his benefactor, he was appointed to the newly created position of Reader in Geology. By all accounts his lectures were delivered in an ebullient manner and were judged as to be so popular and entertaining by the students of the university that he could charge 2 Guineas to attend them. Henry Acland, as a student, attended Buckland's lectures and described his lecturing style thus: *"He paced like a Franciscan preacher up and down behind a long showcase ... He had in his hand a huge hyaena's skull. He suddenly dashed down the steps - rushed skull in hand at the first undergraduate on the front bench and shouted 'What rules the world?' The youth, terrified, threw himself against the next back seat, and answered not a word. He rushed then on to me, pointing the hyaena full in my face - 'What rules the world?' 'Haven't an idea', I said. 'The stomach, sir', he cried (again mounting the rostrum) 'rules the world. The great ones eat the less, the less the lesser still.'"*

In his inaugural lecture as Reader in Geology he spoke of how geology could be reconciled with the Bible and offered nine 'proofs' that the world had been overwhelmed by catastrophic flooding which at the time he correlated with Noah's Flood. His evidence included fossil fauna from caves, superficial deposits of sediments and rocks transported far from their source – erratics. Over time his position of the origin of such features would shift and he developed a more subtle reconciliation between the Bible and the geological record, ultimately viewing Earth history as a series of separate creations.

His first book published in 1823 was *Reliquiae Diluvianae* ('Relics of the Flood') which focused on his excavations at Kirkdale Cave in Yorkshire, where he found an assemblage of bones belonging to extinct species such as mammoth and wooly rhinoceros, as well as animals such as hyena that no longer occur in Europe. In a brilliant study of forensic paleontology he showed that the bones had been accumulated by hyenas that used the cave as a den. He reconciled this with his flood hypothesis by noting that the bones were buried in waterlain silt which he attributed to the Great Deluge and which led to the extinction of some species.

Also in 1823 Buckland discovered a red-stained skeleton in Paviland Cave in South Wales which he named the Red Lady of Paviland. First supposing it to be the remains of a local prostitute he then noted that it occurred in the same strata as the bones of extinct mammals (including mammoth), Buckland shared the view of Georges Cuvier that no humans had coexisted with any extinct animals, and he attributed the skeleton's presence there to a grave having been dug in historical times, possibly by the same people who had constructed some nearby pre-Roman fortifications, into the older layers. In fact, although it post-dates the mammoth remains, it is the oldest anatomically modern human found in the United Kingdom. Carbon-data tests have since dated the skeleton, now known to be male, as from circa 33,000 years before present (BP).

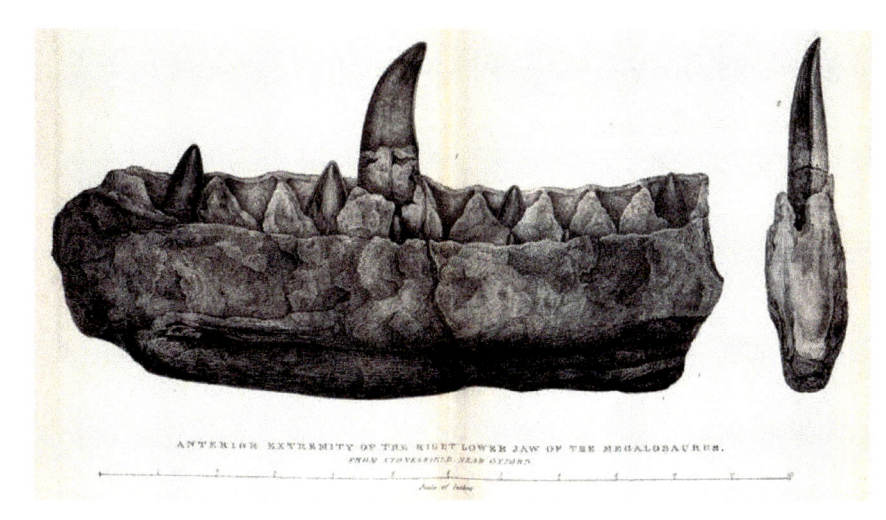

Original illustration of the *Megalosaurus* jaw by Buckland's wife Mary.

Through the 1820's Buckland began to realize that the fossil record suggested a more complex view of Earth history than a simple relationship with the great flood. Following the evidence documented by William Smith, Georges Cuvier and others he realized that there was a stratigraphic organization to the fossil record which contained many extraordinary creatures that have no direct counterpart today. This included the fossil ichthyosaurs and plesiosaurs being found by Mary Anning on the Dorset coast and in 1824 his own analysis of the jaw of what he considered an extinct giant lizard – *Megalosaurus*, later classified as a dinosaur by Richard Owen. This was found by workmen at a quarry in Stonesfield, not too distant from his college rooms in central Oxford. By 1836 he had recognized that the fossil record represented a succession of distinctive past ecosystems and summarized this in his seminal book

Idealised geological cross-section across Europe produced by Buckland in 1836 (in Geology and Mineralogy considered with reference to Natural Theology) illustrating the general succession of strata and their constituent assemblages of fossils. Igneous and metamorphic activity is also highlighted.

William Buckland lecturing at Oxford University.

Geology and Mineralogy considered with reference to Natural Theology. This important book paved the way for others to think about evolution and the stratigraphic organization of the rock record.

As Buckland was writing his geological opus, he became aware of the work of Louis Agassiz on the importance of glaciation. The two soon became collaborators carrying out joint field work in Britain and Switzerland, and Buckland soon realized that the action of past ice ages could explain many of the phenomena that he had previously ascribed to Noah's Flood. For example, the mud covering the bones in Kirkland Cave could be ascribed to sedimentation by glacial meltwaters.

No biographic summary of William Buckland would be complete without mention of some of his noted eccentricities. He often carried out field work in formal academic dress; his house was part home and part free-ranging zoological collection; and he prided himself on having dined on most of the animal kingdom. Mole was declared to be particularly unpleasant. Pausing before a dark stain on the flagstones of an Italian cathedral where a martyr's blood was said to miraculously renew itself, he got to his knees, licked it and declared *"I can tell you what it is: it is bat's urine."* His wife Mary was a supportive collaborator in his studies and assisted in drawing field sketches and experiments at their home. When trying to unravel the origin of the trace fossil *Cheirotherium*, he woke her at 2 a.m. to declare *"My dear I believe the Cheirotherium's footsteps are undoubtedly testitudinal"*. They quickly mixed paste and set their pet tortoise to walk in it, the resultant footsteps matching those of the fossil. Their son Frank would inherit these types of eccentricities when he became a fellow at Oxford, not least by keeping a live bear in his rooms.

In later life Buckland's scientific productivity was limited by his appointment as Dean of Westminster Abbey and by head injuries sustained when his carriage overturned when carrying out field work in France. But he had accomplished a great deal in popularizing geology, making it a university discipline, pushing at the boundaries of the understanding of Earth history and being open-minded enough to modify his interpretations when the evidence demanded it. For those interested in his life and work the excellent Natural History Museum of the University of Oxford has a number of informative displays of some of his achievements.

REFERENCES

This essay has drawn upon information from the following sources:

Annan, N. 1999. *The Dons*. HarperCollins. 357pp.

Cadbury, D. 2000. *The Dinosaur Hunters*. Fourth Estate Ltd, 374pp.

Cook, J. 2007. William Buckland: first to describe a dinosaur. In: Huxley, R. (ed.) *The Great Naturalists*. Thames & Hudson, 241-245.

Donaldson, W. 2002. *Brewer's Rogues, Villains and Eccentrics*. Pheonix, 686pp.

Emling, S. 2009. *The Fossil Hunter*. St Martin's Griffin, New York. 234pp.

Gohau, G. 1990. *A History of Geology*. Rutgers University Press. 259pp.

Hallam, A. 1983. *Great Geological Controversies*. Oxford University Press, 244pp.

Rudwick, M.J.S. 2005. *Bursting the Limits of Time*. The University of Chicago Press, 708pp.

Rudwick, M.J.S. 2014. *Earth's Deep History*. The University of Chicago Press, 360pp.

http://www.oum.ox.ac.uk/learning/pdfs/buckland.pdf

Trinity College, Cambridge University.

Adam Sedgwick

Upon his death in 1728, the physician, naturalist and geologist John Woodward donated part of his extensive collection of fossils to the University of Cambridge. He made an associated bequest to found the Woodwardian Professorship of Fossils (later of Geology). The holder of this post, in addition to curating the fossils, was required to read at least four lectures a year on the general topic of the natural history of the Earth. For this, a salary of £100 a year was paid and the incumbent was required to remain celibate "*lest the care of a wife and children should take the Lecturer too much from study, and the care of the Lecture*".

In 1818, the recently ordained Cambridge graduate and Fellow in mathematics, Adam Sedgwick, was elected to the Woodwardian Professorship. By his own admission, Sedgwick was not a geologist at the time of his appointment. He is said to have remarked, "*Hitherto I have never turned a stone; henceforth I will leave no stone unturned*". He was true to his word. Very rapidly, he established himself alongside Buckland, Lyell and Murchison as one of the greats of the heroic age of geology.

Sedgwick is remembered most for his contribution to developing our understanding of stratigraphy. Working with Murchison, he introduced the Devonian and Cambrian periods into the geological timescale. He was also a brilliant lecturer and can be credited with inspiring Charles Darwin's geological interests. Although he never accepted Darwin's views on evolution, he was active in rejecting literal interpretations of the bible and explaining that geological theories provide a better interpretation of the history of the Earth.

Sedgwick was born on March 22, 1785, the third of seven children of an Anglican vicar, in the picturesque village of Dent, Yorkshire (now Cumbria). Despite his family's modest means, Sedgwick attended nearby Sedbergh School, and then entered Trinity College at Cambridge University on a scholarship. Although he was not a geologist at the time of his appointment to the Woodwardian Professorship, Sedgwick was already noted as a talented academic. He was 5th Wrangler in his graduate year (1808) Mathematics exam (i.e. obtained the 5th highest marks). This led to him being appointed a Fellow at the university with modest teaching duties.

Soon after his appointment to the Woodwardian Professorship, Sedgwick was elected a Fellow of the Geological Society of London. Armed with enthusiasm inspired by the Geological Society's meetings, Sedgwick began undertaking field work to collect specimens to supplement the Woodwardian Collection. He quickly became involved in the great British geological enterprise of the early 19th century – the classification of the stratigraphic record into distinct time periods. By 1822, he was carrying out field work in the Lake District and beginning a long association with the geology of Paleozoic rocks. This led to a friendship with the poet William Wordsworth. Sedgwick

Adam Sedgwick in 1832, aged 47, from a painting by Thomas Phillips.

contributed to Wordworth's *"A Complete Guide to the English Lakes"* with a section entitled *"The geology of the Lake District in four letters addressed to W. Wordsworth, Esq."*

Sedgwick formed an alliance with one of the younger fellows of the Geological Society, Roderick Murchison. From 1827 to around 1840, they were a prolific partnership that created the basis for understanding much of Paleozoic stratigraphy. Unfortunately, it was a partnership destined to end in acrimony and a complete severing of their friendship. Their first field trip, in 1827, visited the western and northern coasts of Scotland (mostly by boat and in a fairly superficial manner), the main aim being to understand the stratigraphic position of the red sandstones that formed significant portions of the outcrop.

By 1831, Sedgwick had begun field work in North Wales (he was *"burnt as brown as a pack-saddle, and a little thin from excessive fatigue"* he wrote to a friend), examining the region's slaty, largely unfossiliferous rocks that he understood to be very old. Murchison placed great emphasis on fossil content; but, perhaps due to his mathematical background, Sedgwick was equally content developing an understanding of the structure of a region. This was to prove particularly useful in determining the stratigraphic position of much of the North Wales succession.

In 1834, he undertook what was to prove a definitive joint field trip with Murchison to the Welsh Borderlands. The result was a seminal statement on their Paleozoic geology: *"On the Silurian and Cambrian Systems, Exhibiting the Order in which the Older Sedimentary Strata Succeeded Each Other in England and Wales"* (read in 1835 and published in 1836). The term Cambrian was introduced for Sedgwick's slates of North Wales, and Silurian for the younger fossiliferous rocks upon which Murchison had focused much of his attention.

Ultimately, the boundary between the two systems was to prove divisive. Murchison (see seperate biography) possessed a vainglorious temperament. Styled by his admirers the "King of Siluria", he sought to expand his Silurian "kingdom", ultimately wishing to include rocks containing the oldest fossils. This included rocks that Sedgwick was adamant should be classified as Cambrian. Murchison persuaded the Geological Survey to colour much of their map of Wales as Silurian. This was too much for Sedgwick. He compared himself to a man who comes home to find *"that a neighbour has turned out his furniture, taken possession, and locked the door upon him"*. Murchison had *"Silurianized the map of Wales"*. The partnership was broken and the controversy only resolved by Charles Lapworth after the death of both Murchison and Sedgwick, when he introduced the Ordovician System, placed between Cambrian and Silurian.

Before the schism between Murchison and Sedgwick, they collaborated on one important project that led to the creation of the Devonian system in 1839. Controversy existed regarding the stratigraphic position and nature of the "Grauwackes" present in the county of Devon in south-west England, that lay beneath Carboniferous limestones. Murchison's initial position was to claim these for his Silurian System, but paleontological evidence suggested something more akin to Carboniferous strata. This dilemma was eventually resolved through the study of sections in Europe, especially in the Rhineland area, which indicated that these rocks represented a correlative with the Old Red Sandstone that warranted a separate stratigraphic term. Accordingly, "Devonian" was introduced in a joint paper by Sedgwick and Murchison in 1839.

Sedgwick served as president of the Geological Society from 1829 to 1831. His presidency coincided with the publication of *Principals of Geology* by Charles Lyell, a book that did much to both popularise geology and provide an embracing theory to explain geological observations.

Sedgwick was not taken with the uniformitarianism of Lyell. In his heart, he supported the more catastrophist views of Baron Georges Cuvier. He was greatly impressed by the work of the French geologist Jean-Baptiste Elie de Beaumont, whose recognition of angular unconformities in the mountain ranges of Europe seemed to provide evidence of sudden cataclysmic events in geological history. In his 1831 presidential address, whilst congratulating Lyell on his magnificent opus, Sedgwick did not hold back from criticising how, in the style of an advocate (which indeed Lyell was!), Lyell had sought observations to support his theory, rather than *vice versa*.

Early in his career, Sedgwick was a "diluvianist", interpreting widespread boulder clays as the result of deposition from the great flood of the bible. By 1831, Sedgwick had seen enough evidence from across Britain and Europe that these deposits were not synchronous in age and represented multiple depositional events. He happily stated this in his Geological Society presidential address, a noteworthy recantation that demonstrated clerical geologists were not bound to literal biblical interpretations to explain the geological observations they made.

Thus, he became increasingly opposed to the scriptural geologists of his time who interpreted geological phenomena as expressions of biblical history. The debates between the two parties could be quite bitter. William Cockburn, the Dean of York Cathedral and a leading scriptural geologist, did nothing to quell the animosity between the two sides when he titled his 1849 book *A New System of Geology: Dedicated to Professor Sedgwick*. Cockburn was still stung by a debate in 1844, at a meeting of the British Association for the Advancement of Science, where Sedgwick defended modern geology, which Cockburn described as "unscriptural." The proceedings of this acrimonious event were reported in the national press, marking a key moment in the conflict between scripture and science.

Whilst Sedgwick envisaged a very ancient Earth (i.e. many millions of years old), he was a staunch opponent of evolution, fearing that it would "*undermine the whole moral and social fabric of society.*" How ironic that it was Sedgwick who was selected to give Darwin a crash course in geology before he set sail on HMS *Beagle* and whose solid grounding in making geological observations led Darwin to develop his ideas of natural selection! Sedgwick was appalled when *On the Origin of Species* was published. He wrote to Darwin: "*I have read your book with more pain than pleasure. Parts I laughed at till my sides were sore; others I read with absolute sorrow, because I think them utterly false and grievously mischievous*". Despite the harsh language, the two remained on friendly terms for the rest of Sedgwick's life.

The village of Dent, Sedgwick's birthplace. In the right foreground is a large granite monument in his name. Dent is surrounded by spectacular scenery, mostly reflecting cyclic deposition of Carboniferous sediments. However, it was not until 1830 that Sedgwick described the geology of his home town region.

At Cambridge, Sedgwick did not neglect the specific obligations of his post as Woodwardian Professor. He ensured that the entire collection of John Woodward was purchased (almost 10,000 specimens in total) and began expanding the collection, including the purchase of ichthyosaur skeletons from Mary Anning. These materials would form the core of a new museum at Cambridge, opened in 1904 and named in Sedgwick's honour after his death. Today, the museum houses a collection approaching two million specimens of rocks, fossils and minerals and includes specimens collected by Charles Darwin during his voyage aboard the *Beagle*.

Sedgwick lived to reach 87 and continued his celebrated lectures at the university until he was 85. As well as the museum in his name, his legacy also lives on in the Sedgwick Club, the oldest student-run geological society in the world. His lecture notes read, "*I cannot promise to teach you all geology; I can only fire your imaginations*".

REFERENCES

This essay has drawn upon information from the following sources:

Clark, J.W. & Hughes, T.M. 1890. *The Life and Letters of the Reverend Adam Sedgwick*. Cambridge University Press.

Hallam, A. 1983. *Great Geological Controversies*. Oxford University Press, 244pp.

Park, C. 2017. *Wedded to the Rocks*. Chris Park, 329pp.

Rudwick, M.J.S. 1972. *The Meaning of Fossils*. The University of Chicago Press, 287pp.

Secord, J.A. 1986. *Controversy in Victorian Geology: The Cambrian-Silurian Dispute*. Princeton University Press, 363pp.

Speakman, C. 1982. *Adam Sedgwick*. The Broad Oak Press, 145pp.

Sir Roderick Murchison

Sir Roderick Impey Murchison, 1st Baronet by Stephen Pearce. A portrait from 1856.

The industrial West Midlands of England may appear at first glance unpromising territory for field work, but are in fact the location for some important Paleozoic outcrops. These include the former quarry called Wren's Nest and associated caverns in Wenlock (Silurian) Limestone near Dudley. A remarkable spectacle occurred here in 1849. According to contemporary reports, 15,000 people turned out to hear Roderick Murchison give a geological lecture in the candlelit caverns, which was followed by a procession to the top of the Wren's Nest site where the Bishop of Oxford, with an appropriate degree of mock ceremony, proceeded to crown Murchison "King of Siluria" in front of the cheering crowd. Murchison was at the time one of the titans of British, indeed international, geology and his recognition was well-deserved. Indeed it is doubtful if any geologist has received so many honours as Murchison, both real and those given in well-meant jest.

In his youth Murchison showed little sign that he would become one of the geological greats. He was born in northern Scotland in 1792 to a land-owning family and received a typical schooling for the gentry of the time in Durham. He was no more than an average student and initially chose a military career serving in the Peninsula War.

By 1815 he had decided against a soldier's career and lived on the income of his family estates to pursue a life that largely revolved around shooting and fox-hunting. Two events were to spark his interest in geology. First was his marriage. His wife Charlotte encouraged him to take an interest in culture and a European tour including visits to the Alps may have stirred his interest in the rocks he saw and their formation. She was a trained artist and her skilled field sketches were to illustrate many of Murchison's publications. The second event was a meeting in 1823 with Sir Humphrey Davy, the celebrated chemist, during a shooting party. Davy encouraged him to attend scientific lectures in London (a fashionable pursuit for gentlemen of the age) and Murchison struck upon those of The Geological Society as being of particular interest. The extent to which geology resonated with him is that only seven years after joining The Geological Society in 1824 he was elected its President (he read his first paper to the society in December 1825). The reason for his advancement was no doubt in part due to his energy, organisational abilities and his connections in society, but it was also due to his natural skill in the science and his enthusiasm for fieldwork and documentation of the results. His early works included papers on the geology of regions of southern England and an early attempt to unravel the structural geology of the Alps.

By the 1830's, the pioneering work of William Smith and his contemporaries was being built upon and the paleontological-based subdivisions of geological time were being recognised as formal systems. It was this "enterprise of stratigraphy" that The Geological Society of London focussed upon, aimed at bringing an order and nomenclature to the diverse rocks of Britain and the wider world.

The subdivision and correlation of that which we now term older Paleozoic remained perplexing. These often poorly fossiliferous, slightly metamorphosed, sediments as present beneath the Old Red Sandstone in Wales had been mapped as "Transition rocks" or using the German miners term "Grauwacke" (greywacke) and their stratigraphic subdivision, correlation and mapping was problematic. It was in the unravelling of the stratigraphy of this deep time that Murchison excelled in and the periods Silurian, Devonian and Permian were first recognised by him, although as we shall see, not without controversy.

In 1831, Murchison set about examining outcrops thought to lie below the Old Red Sandstone in the Welsh Borderlands.

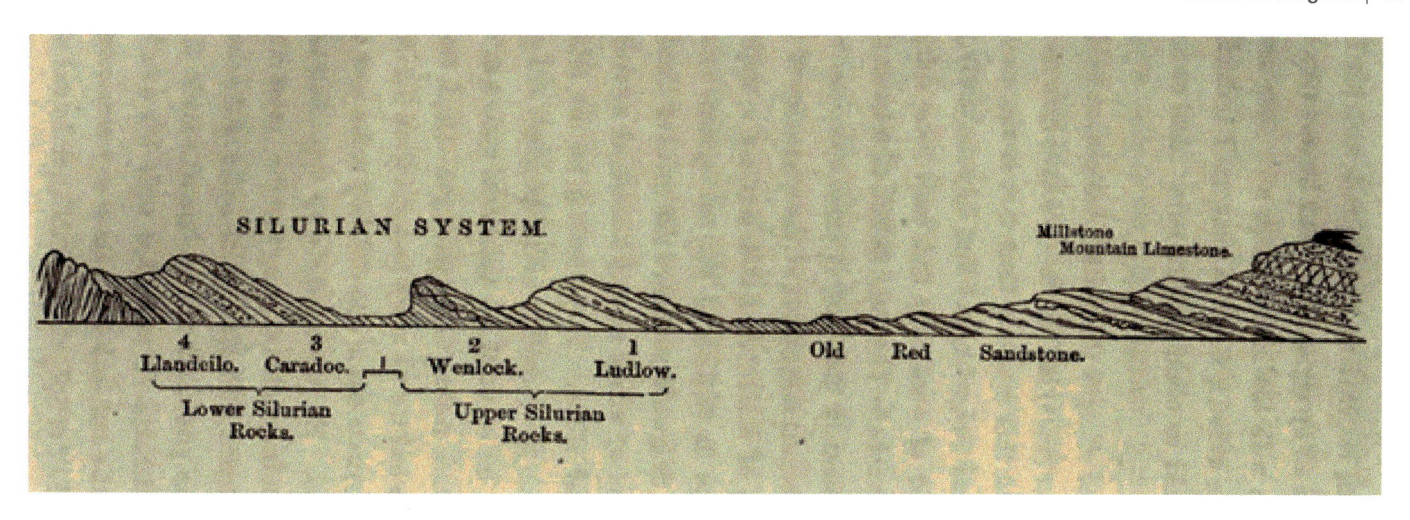

Geological cross-section in the Welsh borderlands from Murchison's paper introducing the Silurian system in 1835. The "Lower Silurian Rocks" were the subject of controversy and are now regarded as Ordovician.

Noting their trilobite and brachiopod faunas, Murchison introduced in 1835 the term "Silurian" (named after the Silures, a Celtic tribe indigenous to Wales that had resisted Roman invasion) for these rocks that lay above the poorly fossiliferous "Cambrian" rocks that had been introduced by his Cambridge-based collaborator, Adam Sedgwick. This included terms such as Ludlow Beds and Wenlock Limestone, terms that are still used today and form the basis for subdivisions of Silurian time.

In 1839 he published his *magnum opus* – "*The Silurian System*", later revised and republished a number of times as "*Siluria*". This assured his fame. It was considered by Leonard Horner, a President of the Geological Society of London *"so accurate in its details, that a very competent judge, who had trod, hammer in hand, over every part of the region, holds it to be the best piece of topographical geology in our language."* Thus in 1840 he was invited by Tsar Nicholas I to carry out field work and report on the mineral wealth of Russia, especially the Urals region. This led to the introduction of the Permian period to describe the distinctive sediments lying above the Carboniferous Coal Measures around the Russian city of Perm. He was also made Knight Grand Cross of St. Stanislaus by Tsar Nicholas in return for his efforts.

By all accounts Murchison was a man confident in his own abilities, possessing a desire to be the centre of attention in the geoscientific world and this led him into a number of conflicts that characterised British geology in the middle of the 19th century.

The first controversy was an understanding of the stratigraphic position and nature of the "Grauwackes" present in the county of Devon in south-west England that lay beneath Carboniferous limestones. Murchison's initial position was to claim these for his Silurian System, but paleontological evidence suggested something more akin to Carboniferous strata. This dilemma was eventually resolved through study of sections in Europe, especially in the Rhineland area, that indicated that these rocks represented a correlative with the Old Red Sandstone that warranted a separate stratigraphic term. Accordingly "Devonian" was introduced in a joint paper by Sedgwick and Murchison. But the two collaborators were to fall out over where the lower limits of the Silurian should lie.

In the 1830's Sedgwick had carried out field work in North Wales and had recognised a series of mostly poorly fossiliferous slates above basement. These he termed "Cambrian". Working at the same time in the Welsh Borderlands and South Wales, Murchison recognised the fossiliferous limestones and shales below the Old Red Sandstone that he termed "Silurian", but difficulty emerged as to which period some of the then oldest fossiliferous rocks in Wales called the Caradoc Sandstones and Llandeilo Flags should belong to. In essence, mistakes and failure to collaborate meant that the "Upper Cambrian" in North Wales turned out to correlate with the "Lower Silurian" in Mid and South Wales. Murchison thus claimed the Upper Cambrian as Silurian (in an effort to include the oldest fossil-bearing rocks in the Silurian), whilst Sedgwick claimed the Lower Silurian as Cambrian. Sedgwick also recognised that there was an unconformity between the "Lower" and "Upper" Silurian and that the term Silurian could be not used beneath this break. Murchison insisted there was no unconformity. A flavour of the acrimonious debate is given in this description by Andrew Ramsay of the presentation of a polemic paper by Sedgwick in February 1852 to The Geological Society of London. *"Good scrimmage between Sedgwick and Murchison on the Lower Silurian and Cambrian question. It was not an enlivening spectacle. Sedgwick used very hard words"*. Matters were not truly resolved until a few years after the deaths of Murchison and Sedgwick, when Charles Lapworth placed the problematic strata within his new Ordovician System based on their graptolite faunas.

A final controversy engaged Murchison in his later years, when, failing to recognise their position above the Moine Thrust, he argued that a series of ancient gneiss, now known to be Precambrian, could not be older than Silurian.

It would be unfair to characterise Murchison as the villain in these controversies and that he was always on the wrong side of the scientific arguments. He was committed to defining an ordered and mappable stratigraphy (in 1855 he was made Director General of the Geological Survey) and indeed controversy persists even today on where the precise limits of geological periods should be drawn. That he tried to resolve these issues through field work, biostratigraphy and structural relationships is to his credit. That he tried to browbeat those who opposed him is not.

Murchison was perhaps the ultimate "gentleman geologist". His position in society allowed him to converse with governments and nobility and there is little doubt that he saw the advancement of geological and geographical knowledge as advancing the British Empire (he was a founder of the Royal Geographical Society and served as its President on multiple occasions). For his efforts he was knighted in 1846 and made a Baronet in 1866. The list of honours he received from both British and international scientific societies are numerous. Nineteen stars, crosses and other emblems of distinction were awarded to him by sovereigns of many nations. Before his death in 1871 he endowed a Chair in Geology at Edinburgh University and a medal and research fund at The Geological Society. His legacy lives on too in many fossils that bear his name (e.g. *Didymograptus murchisoni*, the "tuning-fork" graptolite, a favourite of paleontology students everywhere) and in the name of geographical features such as the Murchison River in Western Australia. He was truly the King of Siluria.

REFERENCES

This essay has drawn upon information from the following sources:

Gohau, G. 1990. *A History of Geology*. Rutgers University Press. 259pp.

Hallam, A. 1983. *Great Geological Controversies*. Oxford University Press, 244pp.

Morton, J.L. 2004. *King of Siluria*. Brocken Spectre Publishing 280pp.

Oldroyd, D.R. 1990. *The Highlands Controversy*. The University of Chicago Press. 438pp.

Rider, M. 2005. *Hutton's Arse*. Rider-French Consulting Ltd. 214pp.

Rudwick, M.J.S. 1972. *The Meaning of Fossils*. The University of Chicago Press, 287pp.

Secord, J.A. 1986. *Controversy in Victorian Geology: The Cambrian-Silurian Dispute*. Princeton University Press, 363pp.

Stafford, R.A. 1989. *Scientist of Empire*. Cambridge University Press.

Fossiliferous Wenlock Limestone (Silurian) from Wren's Nest, Dudley.

Silurian Trilobite (*Calymene blumenbachi*) from Wenlock Limestone, at Wren's Nest. Known colloquially as the 'Dudley Bug', this fossil features in the Coat of Arms of the town of Dudley.

Sir Charles Lyell

As Charles Darwin prepared to sail on the *Beagle* in late December 1831, the ship's captain, Robert FitzRoy, handed him a book that would have a profound influence on the young scientist. That book was the first volume of *Principles of Geology* by Charles Lyell, a man who had made it his personal mission to revolutionise geology such that it would be a science on a par with physics and chemistry. It was not only Darwin that was struck by the theories laid out in the *Principles*, the book was also widely read by both the scientific community at large and by educated members of society in general. It sold widely and eventually ran to eleven editions, the last published in 1875 shortly after Lyell's death.

Charles Lyell was born in 1797 in Kinnordy, north of Dundee in Scotland, and a year after his birth the family moved to Hampshire. His father was a lawyer, and after Charles' graduation in Classics from the University of Oxford in 1819, it was expected that he would also enter the legal profession. However, he had fallen under the spell of geology whilst at Oxford, perhaps inspired by the flamboyant lecturing style of William Buckland, and his career plans lay in that direction. In 1819 he joined the Geological Society of London and his charismatic personality soon saw him elected joint secretary. He would serve as President from 1835–37 and 1849–51.

As a gentleman from a reasonably wealthy family, Lyell was able to indulge in his interests. In a letter to another great geologist of that age, Roderick Murchison, Lyell noted the three things that were essential for success as a geologist: "*travel, travel, travel.*" This was a maxim that he followed all his life, travelling first extensively in Britain and Europe for fieldwork and later in America and Canada. His observations in the field, especially in Europe, led him to consider the gradual, long-term processes that generate the rock record in line with the arguments of James Hutton. Matching the depositional and volcanic processes in operation today with the rock record seemed a much better explanation for Earth history than aligning them with catastrophic events typified by the biblical flood.

Painting of Charles Lyell in 1840 by Alexander Craig.

By 1829 he had begun to develop the key strand of his unifying geological concept: "*…no causes whatever have from the earliest time…to the present, ever acted, but those now acting and that they never acted with different degrees of energy from which that they now exert*" (letter to Murchison, 1829). In other words the geological record could be explained by observable processes acting upon the Earth today at the rate they currently occur, often expressed as "the present is the key to the past." This became known as *uniformitarianism*, a term coined by the Cambridge polymath William Whewell in a review of the second edition of the *Principles*. In contrast, *catastrophism* advocated that the Earth had been subject to successive upheavals and floods, following the ideas of George Cuvier.

Lyell's three volume *Principles of Geology* was published between 1830 and 1833. Its subtitle explains its content and intent: "*an attempt to explain the former changes in the earth's surface by reference to causes now in operation*". Lyell was clear that geology needed a set of laws by which to operate if it was to be judged as a true science. This was the key theme

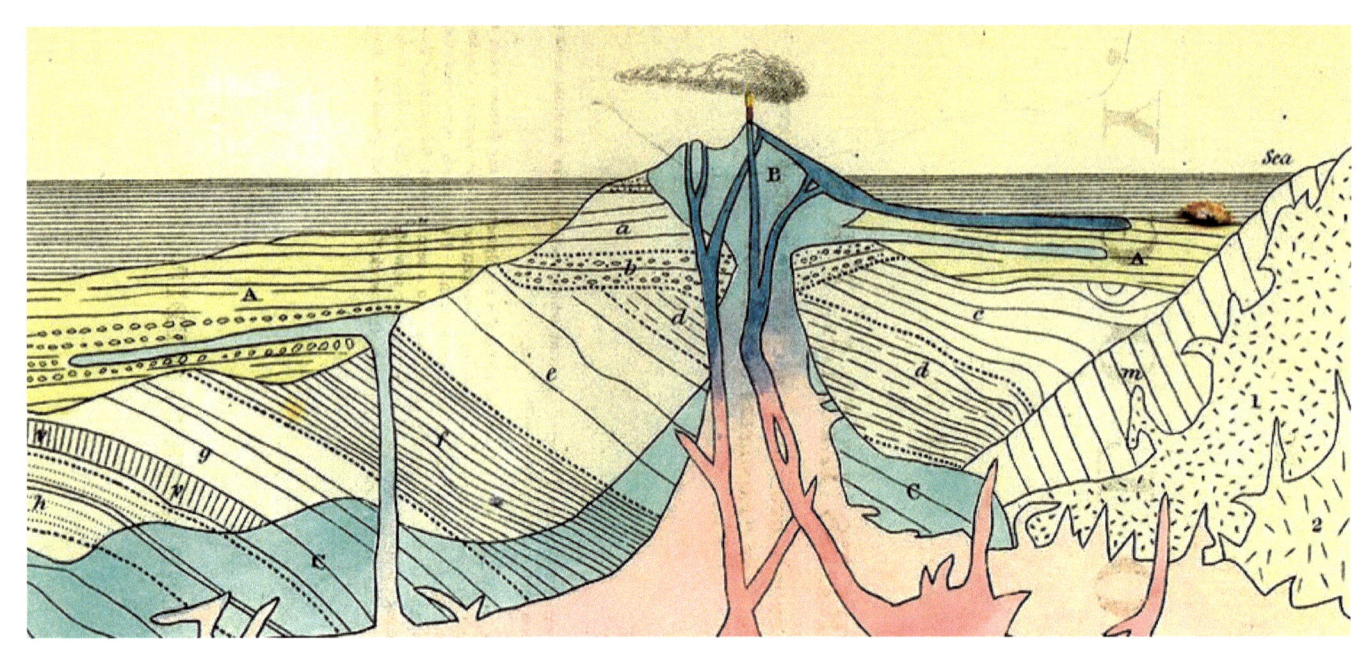

Stratal relationships from *Elements of Geology*, Lyell's synthesis of the *Principles of Geology* aimed at a wider audience.

of the first volume. The second volume was based around refuting Lamarck's ideas on the transmutation of species. Lyell was a devout Christian and viewed the presence of mankind on the planet as exceptional. For '*Ourang-Outangs to become men*' could not be possible. He emphasised the inadequacy of the fossil record and that fossils of modern creatures might well still be found in ancient rocks. Moreover, animals and plants that had once inhabited the Earth might do so again in the future: "*The huge iguanodon might reappear in the woods, the ichthyosaur in the sea, while the pterodactyl might flit again through umbrageous groves of tree-ferns*". Change on Earth was constant and gradual, but led in no particular direction. Nonetheless he was a great believer in the application of fossils for correlation and the definition of periods of geological time and subdivided the Tertiary epoch into the Eocene, Miocene and Pliocene periods based on the relative proportion of modern species present. He also renamed the traditional Primary, Secondary and Tertiary periods (now called eras) to Paleozoic, Mesozoic and Cenozoic, a nomenclature which was gradually accepted worldwide.

Volume three of the *Principles* described the composition of the Earth's crust and phenomena such as volcanoes and earthquakes.

Whilst uniformitarianism became a widely accepted doctrine amongst British and North American geologists in the mid-19th century it was not accepted by all. Geologists such as Adam Sedgwick of Cambridge University argued that folded mountain strata and giant erratic boulders provided good evidence of episodes of "*feverish spasmodic energy*" against which modern processes paled into insignificance. But for Lyell, recourse to such unobserved events required geology to be based on unrestrained speculation — the very thing that in his eyes made the subject unscientific. Better to allow the long reaches of geological time to let modern processes have a cumulative effect and thus provide laws by which geological observations could be interpreted.

Although Lyell and Darwin became friends, Lyell never fully embraced his friends' theory of evolution, despite promoting its presentation in the famous joint paper with Alfred Russell Wallace in 1858. In 1863, Lyell published *The Antiquity of Man* demonstrating the co-existence of humans with now extinct creatures. When Lyell wrote that it remained a profound mystery how the huge gulf between man and beast could be bridged, Darwin wrote "*Oh!*" in the margin of his copy.

We know today that a combination of uniformitarianism and catastrophism is needed to explain the rock record, but Lyell

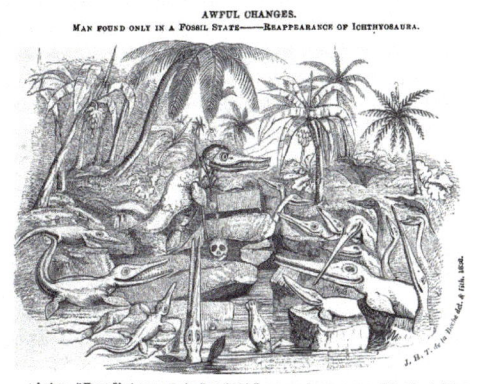

Caricature by Henry De la Beche (1830) ridiculing Lyell's suggestion that Jurassic reptiles, or at least animals like them, might return in a distant post-human future. Not all the contemporary geologists of Lyell fully embraced the implications of his uniformitarianism.

rendered geology a great service by both popularising it, and moving it from being based on theorising and conjecture to linking observations with processes.

Charles Darwin was influenced by reading Lyell's work all his life. He provided this testament to his admiration for Lyell in The Origin of Species in 1859: ...*Sir Charles Lyell's grand work on the Principles of Geology, which the future historian will recognise as having produced a revolution in natural science...*"

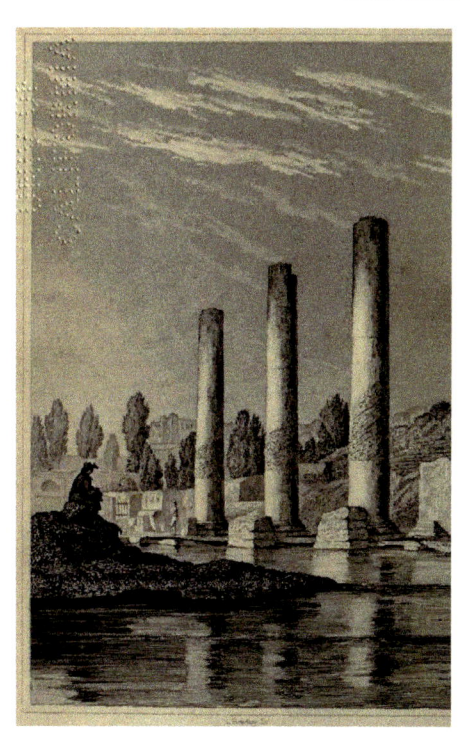

Frontispiece of Charles Lyell's Principles of Geology of 1830, showing the ruined Temple of Serapis near Naples (later re-interpreted as a market building) with its columns marked by a zone in which the stone had been bored by marine molluscs. This showed that in the 2,000 years that had passed since the building had been constructed, this earthquake-prone volcanic region had subsided and risen again – a testament (in Lyell's view) to the steady state dynamic processes of the Earth and the use of the span of human history to explain geological phenomena. It can also be interpreted as documenting catastrophism!

REFERENCES

This essay has drawn upon information from the following sources:

Bonney, T.G. 1895. *Charles Lyell and Modern Geology*. Macmillan & Co, 221pp.

Cook, J. 2007. Charles Lyell: advocate of modern geology. In: Huxley, R. (ed.) *The Great Naturalists*. Thames & Hudson, 246-249.

Gohau, G. 1990. *A History of Geology*. Rutgers University Press. 259pp.

Gould, S.J. 1978. *Ever Since Darwin*. Burnet Books Ltd.

Gould, S.J. 1987. *Time's Arrow, Times's Cycle*. Harvard University Press, 222p.

Greene, M.T. 1982. *Geology in the Nineteenth Century*. Cornell University Press. 324pp.

Hallam, A. 1983. *Great Geological Controversies*. Oxford University Press, 244pp.

Laudan, R. 1987. *From Mineralogy to Geology*. The University of Chicago Press. 278pp.

Oldroyd, D.R. 1996. *Thinking About the Earth*. The Athlone Press, 410pp.

Rudwick, M.J.S. 1972. *The Meaning of Fossils*. The University of Chicago Press, 287pp.

Rudwick, M.J.S. 2014. *Earth's Deep History*. The University of Chicago Press, 360pp.

Secord, J.A. 1986. *Controversy in Victorian Geology: The Cambrian-Silurian Dispute*. Princeton University Press, 363pp.

Wilson, L.G. 1972. *Charles Lyell. The Years to 1841: The Revolution in Geology*. Yale University Press, 553pp.

Highly magnified fossil foraminifera as seen in thin-section

Alcide d'Orbigny

Alcide d'Orbigny was a remarakable prolific and pioneering zoologist, paleontologist, geologist and anthropologist. It was he who first recognised foraminifera as a biological entity and understood their stratigraphic application. As a geologist, he introduced the concept of stages, the now standard means of chronostratigraphic subdivision. Moreover, he related these stages to global events – *"the expression of the boundaries which Nature has drawn with bold strokes across the whole globe"* – immediately bringing to mind modern day concepts of eustatically-driven sequence stratigraphy. But these subjects represent only a small part of d'Orbigny's scientific activity. Several years ahead of Darwin, he undertook a major eight year expedition of South America focused on describing the plants, animals and native people he encountered. On his return, he set about developing an atlas of *all* invertebrate fossils encountered in his native France, along with their associated geology. Sadly, he died at the relatively young age of 55, but not before publishing fundamental works on all his research themes.

d'Orbigny was born in 1802 in the town of Couëron, close to the Loire River, west of Nantes, his family then moving to the coastal village of Esnandes near La Rochelle in 1815 and then La Rochelle itself in 1820. His father was a doctor and a keen amateur naturalist. From an early age, Alcide and his brother Charles accompanied their father on shell-collecting trips on the French Atlantic coast and Alcide began to sketch their finds with a talent for drawing that would subsequently typify his career. Alcide was only 11 years old when he began to examine foraminifera found in the local beach sands, soon supplemented by material sent to his father from Rimini and Corsica and later from around the world. These often beautifully shaped and complex microscopic shells both in living and in fossil form became a subject of special interest to him and by 1825 he was able to use the term Foraminifera for the first time to describe them, although mistakenly regarding them as tiny forms of cephalopod (their true biological affinities, single-celled protists, was determined by Félix Dujardin in 1835). d'Orbigny's *"Tableau méthodique de la classe des Céphalopodes"* presented to the Académie des Sciences in November 1825 enumerated, for the first time, the various taxa he recognized within the foraminifera, along with detailed descriptions of their morphology and mode of life. Both modern and fossil forms were documented and classified. He also made physical models of many species of foraminifera, enlarged 40–200 times, so that their features could be more easily observed. After carving the scaled-up versions out of gypsum, he made plaster replicas which were distributed to universities and museums. Many modern institutions still house collections of d'Orbigny's foraminifera models.

His work soon attracted the interest of the scientific establishment in France which in turn led him being asked to undertake an expedition to South America. Consequently, from June 1826 to March 1834 his travels encompassed the lands from the southern tip of the American continent to the mountains of Bolivia and Peru and much in-between. He collected a vast amount of material and the results of the expedition were published in nine volumes accompanied by around 500 plates. 160 mammals, 860 birds, 115 reptiles, 166 fish, 980 molluscs and zoophytes, 5000 insects and crustaceans and 3000 plants were collected and described, many new to science. The stratigraphical geology of the South American continent was described, supported by palaeontological descriptions. He produced the first geographical and geological maps of Bolivia. His study of the native peoples in South America led to a complete treatise on anthropology. The list of achievements of his participation in this expedition is immense. Nor were his travels without adventure. He passed an unfortunate 15 days of his long trip in a Uruguayan prison (*"a putrid hole full of malefactors and chained murderers"*) after local soldiers found his barometer suspicious.

Alcide d'Orbigny from a lithograph by Lavallée (1839).

The efforts required to document his South American expedition did not deter him from expanding his research into new spheres and in particular the geological record. His first publication on this theme in 1839 was to describe the foraminifera of the Cretaceous Chalk. This included the recognition of fifty new species, but moreover, he surmised that during the Cretaceous period the Paris Basin had been invaded by a warm sea lacking strong currents – a pioneering attempt at using microfossils for paleoenvironmental purposes and to recognize transgression. Following on from this study he undertook the immense task of describing all species of fossil invertebrates found in France in the eight volumes of *"La Paléontologie française"*, a work that remained incomplete at the time of his death.

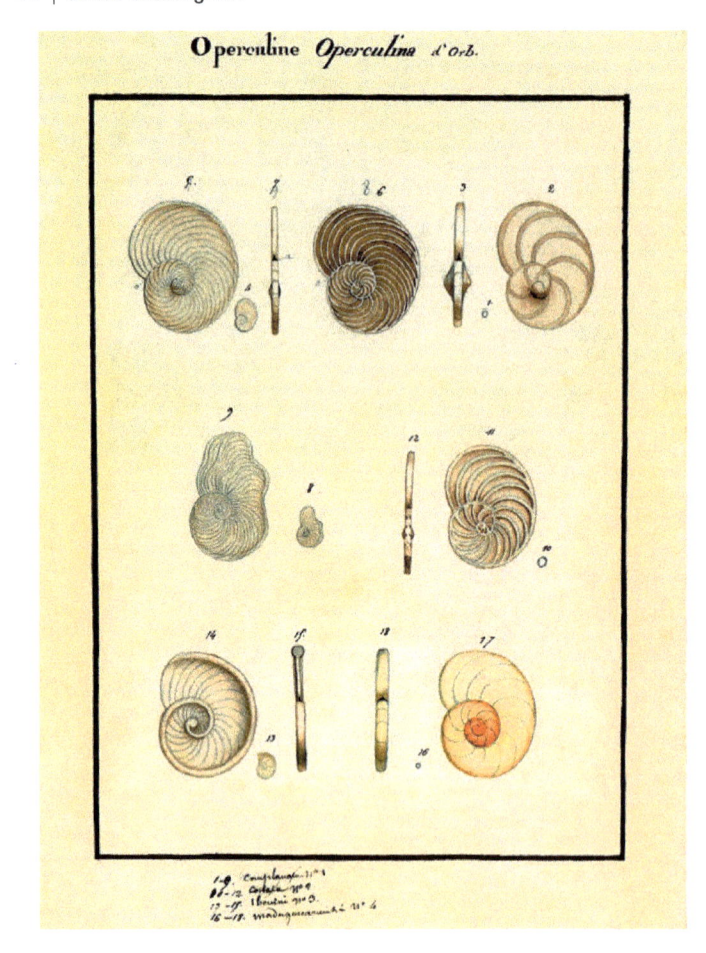

Illustrations of foraminifera by d'Orbigny for the *Tableau méthodique de la classe des Céphalopodes*.

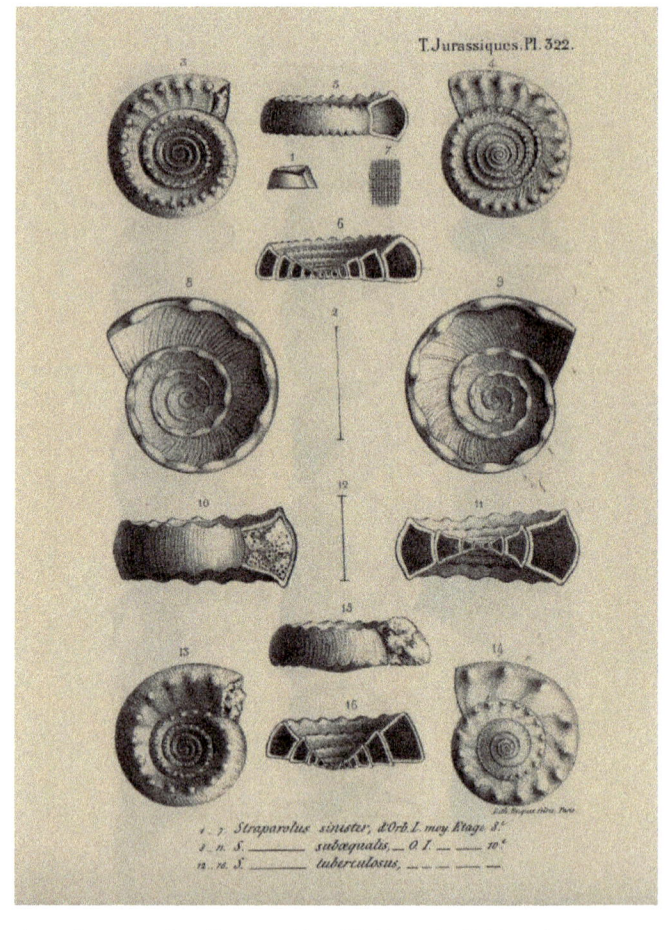

Jurassic ammonites illustrated by d'Orbigny for *Paléontologie Française* (1842).

In 1846 he published his study of fossil foraminifera of the Vienna Basin – a study that like so much of his work was grand in scope and execution. Everything was included: systematics, with the description of 228 species; stratigraphic distribution (Paleozoic to Recent); paleoenvironmental interpretation including possible generalized water temperature variations. In particular, he made an argument for the use of foraminifera in stratigraphic studies and correlation - *"ils peuvent servir à déterminer sûrement l'âge d'un terrain géologique"* (they can be used to determine the age of a geological interval with certainty) - something that is taken for granted today and a vital tool for correlatiing rocks at outcrop and in the subsurface. In the earlier Chalk Memoir he had written *"the comparative study of fossil foraminifera of all zones has proved to me a fact of geological importance: it is that each zone has its characteristic species, by which it may be recognized in whatever circumstances that can occur: and these little shells being infinitely more common than those of Mollusca, the application of them which we can make is so much the more certain and becomes extremely interesting."*

The later years of his life he devoted to his monumental *"Prodrome de Paléontologie stratigraphique des animaux Mollusques rayonnés,"* an attempt to place fossils in stratigraphic context and to *"Cours élémentaire de paléontologie et de géologie stratigraphiques,"* a series of introductory texts on paleontology and stratigraphy. It is within these volumes that d'Orbigny outlined his views on stratigraphic organization, including the introduction of stages (*étages*) as a fundamental concept – bodies of rock characterized by a particular set of fossils and typified by a particular location (a stratotype). In this respect he was building on the ideas of William Smith, but more particularly Baron Georges Cuvier, who explained the extinction of fossil organisms by means of catastrophes. To d'Orbigny each stage (he recognized 27 comprising all of geological history – of course we recognize many more today) was separated by an upper and lower line of demarcation or *"discordance"* representing an episode of natural upheaval. The preserved rocks representing the stages were, in his view, the result of transgression. Thus he viewed stages as transgression separated by unconformities – a forerunner of sequence stratigraphic concepts. We would have to disagree with d'Orbigny in his view that the catastrophes separating each stage were near-absolute – effectively each stage represented, to him, a new creation.

Nonetheless his view of successive transgressions separated by unconformities can be considered the first step on the journey to the sequence stratigraphic models that have been developed and applied in the last few decades. Because d'Orbigny was working mainly on outcrops in platform locations (his stratotypes were "up-systems tract" in a sequence stratigraphic sense) we would anticipate them being bounded by unconformities (i.e. sequence boundaries) with associated faunal turnover.

Posterity has not always been kind to the memory of d'Orbigny's contributions to our science. For many years after his death his proliferation of species names for very similar taxa (*"this wild orgy of nomenclature"* to quote Edward Heron-Allen) and his ultra-catastrophist stratigraphic stance led to a critical view of his work being adopted by zoologists, paleontologists and geologists alike. Subsequently, however, his immense contributions have become recognized; both as the initiator of the science of stratigraphical micropaleontology and as the architect of the cornerstone of chronostratigraphy; the concept of the stage. Moreover, his ideas are now in tune with what we today term event stratigraphy, mass extinctions and sequence stratigraphy. Earth history can indeed be subdivided by global events. As the international stratigraphic community strives to formally define stages as the basic nomenclature of our science, perhaps we will eventually see a harmonization of event stratigraphy, sequence stratigraphy, biostratigraphy and chronostratigraphy – something that no doubt d'Orbigny would approve of.

REFERENCES

This essay has drawn upon information from the following sources:

Gohau, G. 1990. *A History of Geology.* Rutgers University Press. 259pp.

Legré-Zaidline, F. 2002. *Alcide Dessalines d'Orbigny (1802-1857).* L'Harmattan. 249pp.

Monty, C.L.V. 1968. d'Orbigny's concepts of stage and zone. *Journal of Paleontology*, 42, 689-701.

Rioult, M. 1969. Alcide d'Orbigny and the stages of the Jurassic. *The Mercian Geologist*, 3, 1-30.

Torrens, H.S. 2002. From d'Orbigny to the Devonian: some thoughts on the history of the stratotype concept. *Compte Rendus Palevol*, 1, 335-345.

Charles Darwin

There is no doubt that Charles Darwin stands alongside Copernicus, Galileo and Newton as one of the greats of science. His book, *On the Origin of Species*, is one of the most widely known and readable natural history texts and has never been out of print since it was first published in 1859. The insights Darwin gained whilst participating in the voyage of the *Beagle* (from 1831 to 1836) and his gradual, cautious development of the ideas of natural selection, a cornerstone of the theory of evolution, are known to everybody with an interest in science.

Rightly claimed by biologists as one of their greats, can he also be claimed as a "great geologist?" The answer is undoubtedly yes. Natural selection gives substance and explanation to biostratigraphy and is central to any understanding of Earth history. Moreover, Darwin in part trained as a geologist. Geological observations and theories were an integral part of what he learned during the *Beagle* voyage, as documented in 1,383 pages of notes he compiled about geology — compared with a mere 368 pages of notes on plants and animals. He also authored three key publications on geology long before the *Origin* appeared.

Born in Shrewsbury in 1809, to a moderately wealthy family, Darwin was seemingly an indifferent student, both at school and university. "*You care for nothing but shooting, dogs and rat-catching and will be a disgrace to yourself and all your family*," wrote his father. Initially sent to Edinburgh to study medicine and eventually follow in the family practice, he found the curriculum of little interest, although he used his time there to pursue an interest in natural history. Opting not to complete a degree, Darwin moved to Cambridge to study for a career in the church. Priesthood was a congenial profession that would have allowed him to maintain a gentlemanly interest in the natural world around him.

Natural history had been in Darwin's blood from an early age. He may not have been an exceptional student, but he was an enthusiastic collector of insects, plants and rocks. "*When 9 or 10, I distinctly recollect the desire I had of being able to know something about every pebble in front of the hall door – it was my earliest, only geological aspiration at that time,*" he noted in his autobiography.

Although Darwin attended geology lectures at Edinburgh, it was his interaction with Adam Sedgwick at Cambridge that more likely cemented an interest in geology. Sedgwick was one of the best teachers of geology in Britain at that time and, alongside his friend (later his rival) Roderick Murchison, was engaged in unravelling the Paleozoic stratigraphy of Wales and the Welsh borderlands. In the summer of 1831, after he had obtained his BA degree at Cambridge, Darwin joined Sedgwick for a geological excursion to various regions in north Wales, a crash course in geological fieldwork that was to prepare him for the events to come.

Upon returning from Wales, Darwin was offered an opportunity that was to change his life completely and, we might suppose, our understanding of the evolution of life on Earth. He joined the HMS *Beagle* as the gentleman naturalist companion to the ship's captain, Robert FitzRoy. The *Beagle* was about to embark on a circumnavigation of the world for the Navy Hydrographic Office. The role completely suited the compulsive collector of specimens and a man with an "*enlarged curiosity,*" in the words of his uncle.

Darwin left England as a keen amateur zoologist, botanist and geologist, but returned as an expert with an intellectual agenda, first concentrating on geology. During the voyage, Darwin had digested the seminal *Principles of Geology*, by Charles Lyell. Lyell was the champion of uniformitarianism, a view that the processes acting upon the Earth now are the same as in the geological past and could be used to explain the rock record.

This contrasted with the catastrophism that Sedgwick would have taught Darwin at Cambridge. This doctrine supposed that the Earth had been subject to episodic violent geological and biological change. Lyell's view of the Earth as being in a steady state of modification greatly influenced Darwin, and he sought evidence to support it as the *Beagle* travelled the globe. The existence of an Earth subject to gradual change permeated not only his geological theories, but also his development of the concept of natural selection as an explanation for the development of species through time. Darwin's great champion, Thomas Huxley, commented, *"I cannot but believe that Lyell was for others, as for me, the chief agent in smoothing the road for Darwin"*.

During the *Beagle* voyage, Darwin gathered evidence to argue that many features in the landscape could be explained by the slow movement, over millennia, of vast blocks of the Earth's crust see-sawing on a shifting molten substructure, with elevation in one area matched by subsidence in another (note the emphasis on vertical movement, not lateral movement as in plate tectonics). At Saint Jago in the Cape Verde Islands, Darwin carried out his first geological fieldwork of the voyage, which led to the application of uniformitarianism to explain a gradual elevation of the whole island and gradual subsidence around volcanic craters.

After being caught in an earthquake whilst they were surveying the coast of Chile, the *Beagle* crew repeated their earlier measurements and discovered a small change in the relative height of land and sea. Darwin had already suspected a gradual elevation of the land, based on fossils he had collected in the Andes. Shortly before the earthquake, they had witnessed the eruption of an inland volcano, Mount Osorno. Could these events be connected?

Darwin literally pieced together his observations, gluing pieces of paper into long strips to make geological cross-sections of the Andes, demonstrating the deformation of the strata. He recognised that the Andes had undergone progressive, not sudden, elevation and that earthquakes, volcanic activity and elevation of land were interconnected (a theory he presented to the Geological Society in 1838, receiving enthusiastic acclaim, not least from Lyell). Armed with the concept of gradual subsidence and uplift, Darwin went on to investigate further geological problems.

In the second volume of Lyell's work, Darwin read about the supposed formation of coral atolls. Although he had not yet seen one, he devised a new theory of their formation, based on the subsidence of oceanic mountains and the offsetting growth of the fringing coral reefs. When the *Beagle* reached the Cocos (Keeling) Islands in the Indian Ocean, he saw that his theory was correct, tested by dredging the ocean floor. Darwin suggested that coral atolls had begun as reefs encircling islands, which then sunk beneath the water as the landmass beneath them subsided. Corals had continued to grow as subsidence occurred, keeping pace with the position of the photic zone. Thus, initially volcanic islands will be surrounded by fringing reefs, then by barrier reefs detached by sea from land, and finally atolls will appear, consisting of a ring of coral reef with a central shallow lagoon. His theory was substantiated in

Portrait of Charles Darwin by George Richmond. Late 1830s, after Darwin had returned from the voyage of the Beagle.

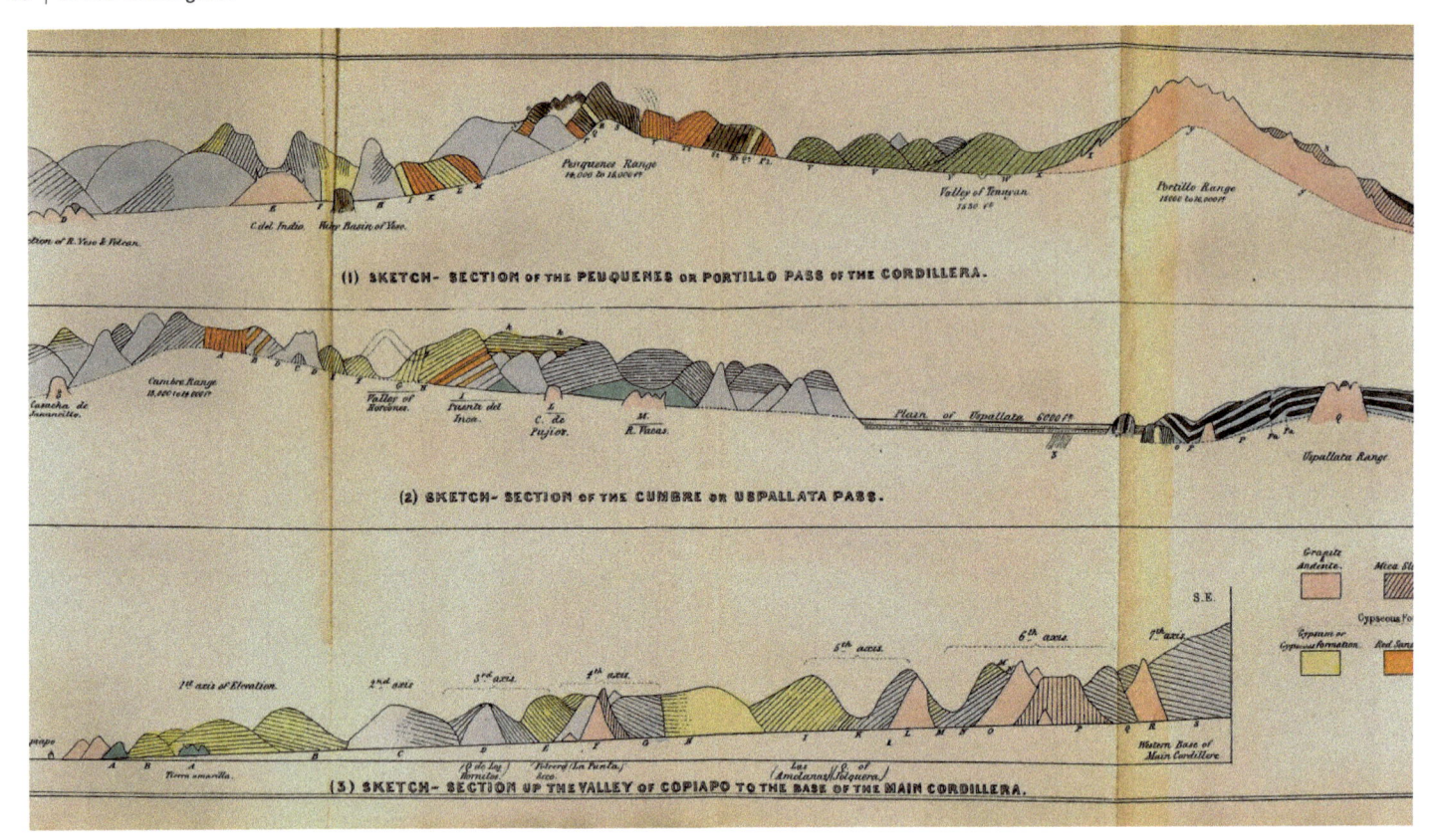

Geological cross-sections of the Andes by Darwin from *Geological Observations on South America* (1846).

1952, when the U.S. Atomic Energy Commission drilled a borehole near the Bikini Atoll in the Pacific Ocean, which showed a coral reef 1,300 m thick sitting directly on bedrock. A sign was placed at the borehole location, announcing "*Darwin was right*".

Darwin thought a similar process might explain the "parallel roads" of Glen Roy in Scotland, a series of striking horizontal terraces. He suggested that these were marine terraces, created by periodic connections with the oceans when the land rose or fell. Soon afterwards, Louis Agassiz stressed the importance of glaciation on the landscape. He demonstrated that the features represented beach terraces alongside a glacially-dammed lake. Darwin described his error as "*one long gigantic blunder*". Nonetheless, the general importance of subsidence versus uplift remains an important geological concept to this day.

Aboard the *Beagle*, Darwin began mulling over the problem of the evolution of species (then termed 'transmutation'). His grandfather, the polymath Erasmus Darwin, had published on the subject in his *Zoonomia* and he was aware that the great French naturalist, Jean-Baptiste Lamarck, had promoted the idea, although neither had produced convincing theories as to how one species might 'transmute' into another. Most scientists of the time favoured divine creation of species, post-catastrophic events that removed the majority, if not all, species from the planet. Darwin preferred gradual rather than catastrophic change. For example, as he collected fossils in South America, he noted that Pleistocene armadillos and sloths, although distinct from modern forms, were close enough to suggest a relationship. What he needed was a mechanism to drive the development of new species. Within nine months of

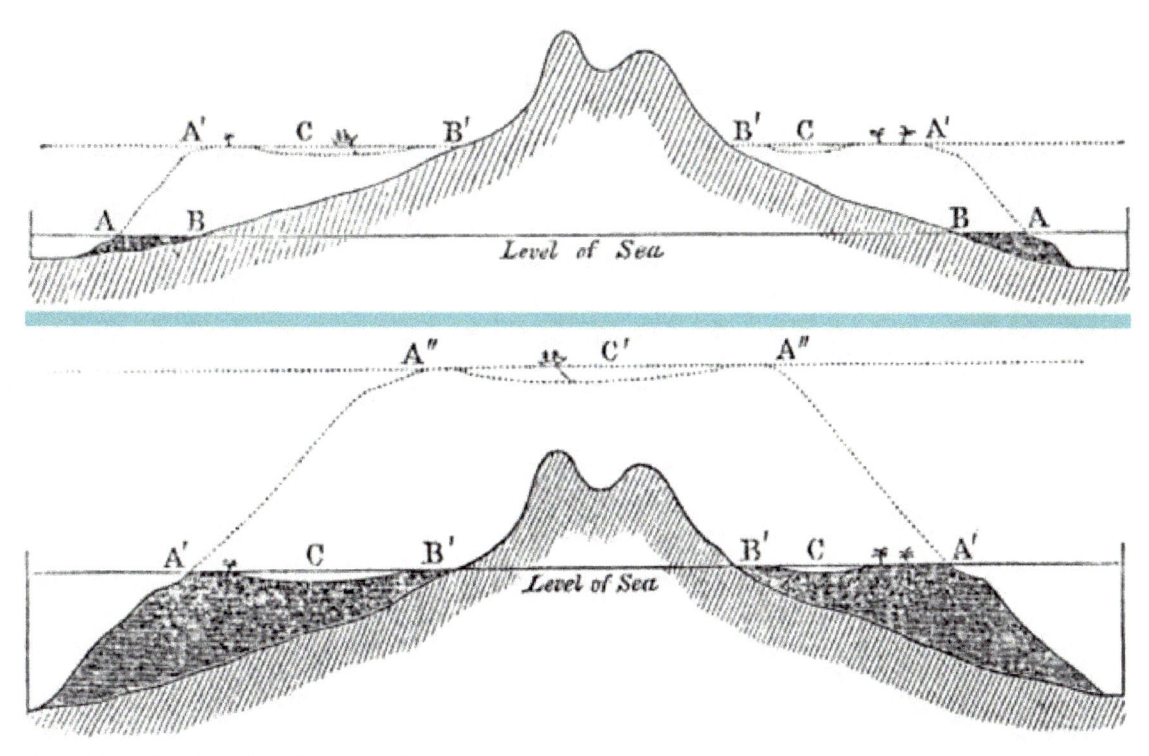

Darwin's explanation of the formation of coral atolls. Top figure: lower solid line, stage 1, a fringing reef (AB) abuts the shoreline. Island sinks (because of subsidence) to upper dotted line, stage 2, barrier reef (A′) separated from sinking island by lagoon (C). Bottom figure: lower solid line, stage 2 barrier reef (as for upper figure). Island sinks further to upper dotted line, stage 3, an atoll (A″), enlarged lagoon (C′) marks previous location of sunken island. From *The Structure and Distribution of Coral Reefs* (1842).

returning home, Darwin started a new notebook dedicated to capturing his developing ideas on evolution, by 1842, a 'first-sketch' of the natural selection theory was complete.

An important development was Darwin's reading of Thomas Malthus' work on population expansion, an intellectual concern in the 19th Century. In his autobiography, Darwin notes, *"In October 1838 … I happened to read for amusement Malthus on Population, and being well prepared to appreciate the struggle for existence that everywhere goes on from the long-continued observation of the habits of animals and plants, it at once struck me that under these circumstances, favourable variations would tend to be preserved and unfavourable ones to be destroyed. The result of this would be the formation of new species"*. This is the essence of natural selection! In a given ecological niche, competition to survive would favour any characteristic that would confer an advantage on offspring. Over time, these characteristics would increase, resulting in a form that was distinctly different from its remote ancestors.

Using his documentation from the *Beagle's* expedition, Darwin established himself in scientific society in London with the publication of *Journal of Researches* (1839). In 1842, needing peace and privacy in which to work and suffering from progressively ill health, Darwin and his wife Emma Wedgwood (his first cousin) moved to Down House in Kent. There, surrounded by a growing family (ten children, seven of whom survived to adulthood), he worked for just two or three hours each day in isolation from the normal cares of the world, but in correspondence with scientists from across the globe on a huge range of topics. Despite this limited work schedule, his output was immense.

Although working on natural selection, Darwin's first major publications from the early days at Down House were geological. A volume on coral reefs appeared in 1842, volcanic islands in 1844 and *Geological Observations on South America* in 1846. His attention then turned to barnacles, both fossil and extant, which occupied him for much of the time prior to the publication of the *Origin* in 1859.

The story of the *Origin's* publication is well known. In 1858, Darwin received a letter from the naturalist Alfred Russell Wallace, working on the Malay Peninsula, that outlined the ideas of natural selection, very much as Darwin had contemplated as early as 1842. Encouraged by Lyell and Huxley, a joint paper by Darwin and Wallace was presented (to no great acclaim) at the Linnaean Society on 1st July 1858. This was followed by the publication of the *Origin* on 25th November 1859, which summarised the huge amount of evidence Darwin had gathered, intending to publish a much more detailed work called *Natural Selection* later (this never happened in his lifetime; it was only pieced together from manuscript form in 1975). Not surprisingly, given the prominence it places on natural rather than divine processes and the implications that man and apes share a common ancestor (not, as many people thought, that we are descended from the apes of today), the *Origin* created a sensation, and was ultimately printed in six editions.

One great strength of the book is that it anticipated and answered the objections of critics. The geological record was imperfect, but the Earth had to be extremely old to provide time for evolution to create a diversity of species and the development of new structures, such as wings from limbs. He even offered up a test: "*If it could be demonstrated that any complex organ existed, which could not possibly have been formed by numerous, successive, slight modifications, my theory would absolutely break down. But I can find no such case*".

The geological observations made during the voyage of the *Beagle* enabled Darwin to grasp two fundaments needed for his scientific theory: 1) the existence of deep time, and 2) the slow, but perpetual changes of the Earth, itself. If geology provided the evidence of an ever-changing Earth over the vastness of time, then organisms had to adapt to survive.

Darwin described himself as having "*unbounded patience in long reflecting over any subject.*" His delay in publishing the ideas that eventually emerged in the Origin is the stuff of scientific legend. Yet the methodical capture of evidence to support his theories and the careful attention to potential criticism made his work all the stronger. He was, thus, recognised as one of the greats of science in his lifetime, although as a modest man he refused most honours. Darwin is buried in Westminster Abbey, arguably the most complete naturalist of all time.

REFERENCES

This essay has drawn upon information from the following sources:

Bynum, W. 2009. Introduction to the Penguin Classics edition of *On the Origin of Species* by Charles Darwin.

Gould, S.J. 1983. *Hen's Teeth and Horse's Toes*. W.W. Norton & Company, Inc., 413pp.

Greene, M.T. 1982. *Geology in the Nineteenth Century*. Cornell University Press. 324pp.

Hallam, A. 1983. *Great Geological Controversies*. Oxford University Press, 244pp.

Herbert, S. 2005. *Charles Darwin, Geologist*. Cornell University Press. 485pp.

Johnson, M.E., Baarli, B.G., Cachão, M., Mayoral, E.., Ramalho, R.S., Santos, A. & da Silva, C.M. 2018. On the rise and fall of oceanic islands: Towards a global theory following the pioneering studies of Charles Darwin and James Dwight Dana. *Earth-Science Reviews,* 180, 17-36.

Oldroyd, D.R. 1996. *Thinking About the Earth*. The Athlone Press, 410pp.

Rudwick, M.J.S. 1972. *The Meaning of Fossils*. The University of Chicago Press, 287pp.

Rudwick, M.J.S. 2014. *Earth's Deep History*. The University of Chicago Press, 360pp.

Thomson, K. 2007. Charles Darwin: the complete naturalist. In: Huxley, R. (ed.) *The Great Naturalists*. Thames & Hudson, 267-276.

Louis Agassiz

Louis Agassiz photographed c. 1865. Photographer unknown.

The Aargletschers (or Aare Glaciers) are a remote, yet undeniably spectacular location in the Bernese Alps of Switzerland. Even today, they require a serious effort to visit and explore. Yet in 1840, a dynamic Swiss scientist had a hut constructed there ("Hôtel des Neuchâtelois") to live in whilst studying the structure and movement of the ice. This scientist was Louis Agassiz, a biologist and paleontologist, who had previously gained fame for his work on living and fossil fish. He emigrated to the United States in 1846, becoming a renowned professor at Harvard University, arguably the first great American scientist. Agassiz has become a controversial figure, not least for his staunch opposition to evolution and his views on race, yet there is no denying his significant contributions to geology, both paleontological and in promoting the view that the Earth had, in its relatively recent past, been subjected to an "ice age".

Agassiz was born in Môtier in the Fribourg canton of Switzerland in 1807. Originally planning for a career in medicine, he studied at the universities of Zurich, Heidelburg and Munich. However, an interest in natural history led him to be selected for the study of a collection of fish brought back from an expedition to the Amazon. All thoughts of medicine were abandoned, and with an enthusiasm that was to characterise all his scientific endeavours, he completed and published the descriptions of this fauna in 1829. This, and an interest in the freshwater fish of Central Europe, led to his appointment as Professor of Natural History at the University of Neuchâtel in 1832.

By this time, fossil fish had come to his attention. Agassiz visited the principal museums of Europe to study fossil fish collections. This included a stay in Paris with Georges Cuvier, the great French paleontologist, geologist and zoologist, whose skills in comparative anatomy must have inspired the young Agassiz. Cuvier was also a strong promoter of a catastrophist view of Earth history and no doubt, this too, influenced Agassiz in his subsequent geological work.

Five volumes of *Recherches sur les poissons fossiles* ("Research on Fossil Fish") were published by Agassiz at intervals from 1833 to 1843, describing and classifying 1,700 species. These works are hardly surpassed today in the quality of their descriptions and illustrations (the latter being mainly undertaken by Joseph Dinkel). The 1,290 original illustrations for the volumes can be found housed in The Geological Society in London. They were donated to the Society by the Earl of Ellesmere, a scientific benefactor, who purchased them to fund Agassiz's research. The sheer scale of these publications, which incorporated a novel new ichthyological classification, marked Agassiz as a leading scientist of the day.

Further paleontological research involved descriptions of fossil echinoderms and molluscs and a special study of the remarkable fossil fish of the Devonian Old Red Sandstone from the Orcadian Basin of northern Scotland. But by 1837, Agassiz's thoughts were turning to another subject — ice and its widespread

The Aare Glacier in the Swiss Bernese Alps, location of some of Agassiz's early research into glacial phenomena.

presence in the geological past. The glaciers of the Swiss Alps were a subject of increasing scientific interest by the late 1830s. Jean de Charpentier and Karl Schimper, and even James Hutton had already arrived at the conclusion that the erratic blocks of Alpine rocks scattered over the slopes of the Jura Mountains had been transported there by glaciers. Agassiz went further. In 1837, he gave a lecture to the Swiss Society of Natural History, in which he envisaged that an ice sheet had smothered the entire northern hemisphere in a prolonged ice age.

The reaction was mostly skeptical — some of his harshest critics suggesting that he should concentrate on fossil fish rather than getting involved in subjects he knew little about. This was not entirely surprising, as Agassiz offered little explanation as to why glacial expansion had occurred. Indeed, at the time, his concept of an ice age could be described as a theory looking for both convincing evidence and a driving mechanism. But Agassiz was undeterred. With typical energy, he conducted research at the "Hôtel des Neuchâtelois" and published a two-volume work in 1840, entitled *Études sur les glaciers* ("Studies on Glaciers"). In it, he discussed the movements of glaciers, the moraines they produce and how evidence for their movement can be seen in the grooves and striations made in rocks over which ice has travelled.

Such observations were brought to bear on the geomorphology of Scotland, Wales and parts of England in a tour Agassiz made with William Buckland in 1840. Agassiz hoped to win over the major name in geology of the time, Charles Lyell. However, the great uniformitarianist was hard to persuade. A phenomenon like an ice age sat ill against Lyell's view that global processes operating today were the same as those in the geological past. Other great geologists of the time also remained cool on the subject of ice ages. Roderick Murchison repeatedly dismissed "ice-mad geologists" in lectures at the Geological Society.

Nonetheless, many others were convinced by the growing evidence for a major widespread glaciation. Buckland made the following entry into the visitors book of the Goat Hotel, Beddgelert in North Wales in 1841: "*Notice to geologists – At Pont-aber-glass-llyn...see a good example of the furrows, flutings and striae on rounded and polished surface of the rock, which Agassiz refers to the action of glaciers. See many similar effects on the left, or south-west, side of the pass of Llanberis*". Suddenly, the evidence for glaciation was being seen where it had previously been ignored. More than just gathering evidence, scientists began to speculate on the wider effects of such glaciation — changing sea-level and isostatic rebound after the glacial ice melted. By the late 19th century, geologists such as

Ctenoid fish fossil, from Louis Agassiz, Recherches sur les poissons fossiles, 1833–43.

T.C. Chamberlin were actively attempting to map the extent of the North American Pleistocene ice sheet. They also recognised, as Agassiz had suggested, that there was more than one period of ice sheet expansion. In short, Agassiz had instigated the science of glaciology and study of glacial phenomena in the geological record.

For much of the early part of the 1840s, Agassiz was involved in compiling his *Nomenclator Zoologicus*, a classified list, with references, of all generic names and those of higher taxonomic groups employed in zoology. By 1846, he was ready for a new challenge and aided by a grant from the King of Prussia, Agassiz crossed the Atlantic with the twin purposes of investigating the natural history and geology of North America and delivering a series of lectures on "The Plan of Creation as shown in the Animal Kingdom". Like Buckland and many other geologists of that time, Agassiz endeavored to square religious teachings with his observations, a trait that would ultimately lead him to take an anti-evolutionary stance.

Society in Boston welcomed this charismatic visitor with open arms and he was soon persuaded to stay, the city and Harvard University becoming his home and scientific base for the rest of his life. He was a gifted teacher, emphasising the study of specimens and outcrops over book-learnt knowledge, and many of his students became notable scientists in their own right. He continued his studies into past glaciations and recognised the sediments of a glacial lake that was a precursor to modern-day Lake Winnipeg (subsequently named Lake Agassiz in his honour). As is demonstrated by the numerous geographic locations and fossil and living taxa named in his honour, Agassiz was arguably one of the best known natural historians in the world in the 1850s. A celebrity scientist in modern terms.

Troubled by ill health, he still took part in two notable expeditions, one to Brazil from April 1865 to August 1866 and a further expedition to South America, travelling through the Magellan Straits in 1872. These trips had many objectives, but one key strand was to gather evidence to refute the tenants of evolution, as laid out by Charles Darwin in *On the Origin of Species,* published in 1859. The idea of the "transmutation of species" was an anathema to Agassiz. In his view, every species was created by design — "a thought of God" — and God had provided different environments through time, in which successive groups of species could flourish before being replaced with a more advanced set. This was creationism meets catastrophism.

Ironically, Agassiz gathered evidence that was used to support evolution. He noted that within a particular taxonomic group,

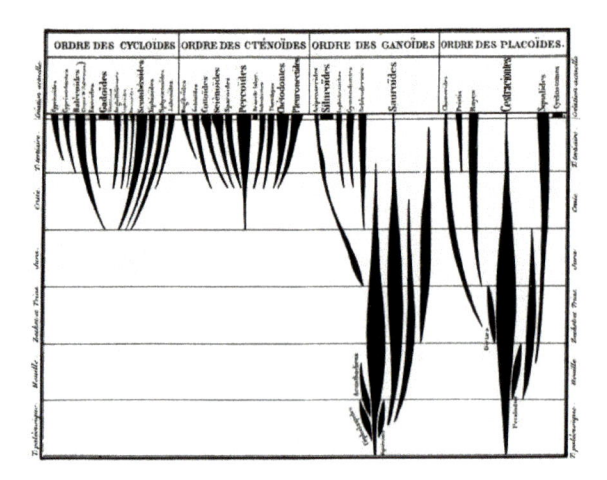

Agassiz's diagram published in 1844 showing the distribution of different fossil fish taxa through geological time.

Rocks polished and striated by a glacier, from *Études sur les glaciers*, 1840.

simpler forms precede more complex forms in the geological record and this development is repeated in embryonic development. Often expressed as "ontogeny follows phylogeny" this is now considered an over-simplification. Nonetheless, Darwin wrote in *On the Origin of the Species* that "*this doctrine of Agassiz accords well with the theory of natural selection*". As early as 1844, Agassiz was publishing charts of fossil fish species distribution through time that could be considered evolutionary trees, but were in his mind simply designed to show that more simple forms preceded more complex forms as designed by God.

Biographical portraits of Agassiz have become increasingly damning in recent years. He is often portrayed as a scientific showman, whose charisma led to success in fund-raising that was the envy of many and whose scientific contributions were often built on the work of others. He eventually became intellectually isolated as demonstrated by his resistance to evolutionary theory. Whilst it is true his ideas on an ice age were developed from Jean de Charpentier and Karl Schimper, who had taken him into the field (Schimper had even mentioned the term "eiszeit" (ice age) to him), it was Agassiz who developed them on a grand scale, gathered evidence and publicised them effectively. His work on fossil fish was unique and he tirelessly promoted science in America, creating the famous Museum for Comparative Zoology at Harvard. We have already noted the irony in that his paleontology and embryological work helped Darwin and others to develop evolution as a theory. He may have been unwitting in this, but the insight he allowed is important nonetheless. Let us remember Agassiz for these contributions.

REFERENCES

This essay has drawn upon information from the following sources:

Gould, S.J. 1980. *The Panda's Thumb*. W.W. Norton & Co., Inc.

Gould, S.J. 1983. *Hen's Teeth and Horse's Toes*. W.W. Norton & Company, Inc., 413pp.

Irmscher, C. 2013. *Louis Agassiz: Creator of American Science*. Houghton Mifflin Harcourt Publishing Company. 434pp.

Lurie, E. 1960. *Louis Agassiz, a Life in Science*. University of Chicago Press.

McPhee, J. 1998. *Annals of the Former World*. Farrar, Straus and Giroux. 696pp.

Rudwick, M.J.S. 2014. *Earth's Deep History*. The University of Chicago Press, 360pp.

Williams, D. 2007. Jean Louis Rodolphe Agassiz: examination, observation, comparison. In: Huxley, R. (ed.) *The Great Naturalists*. Thames & Hudson, 261-266.

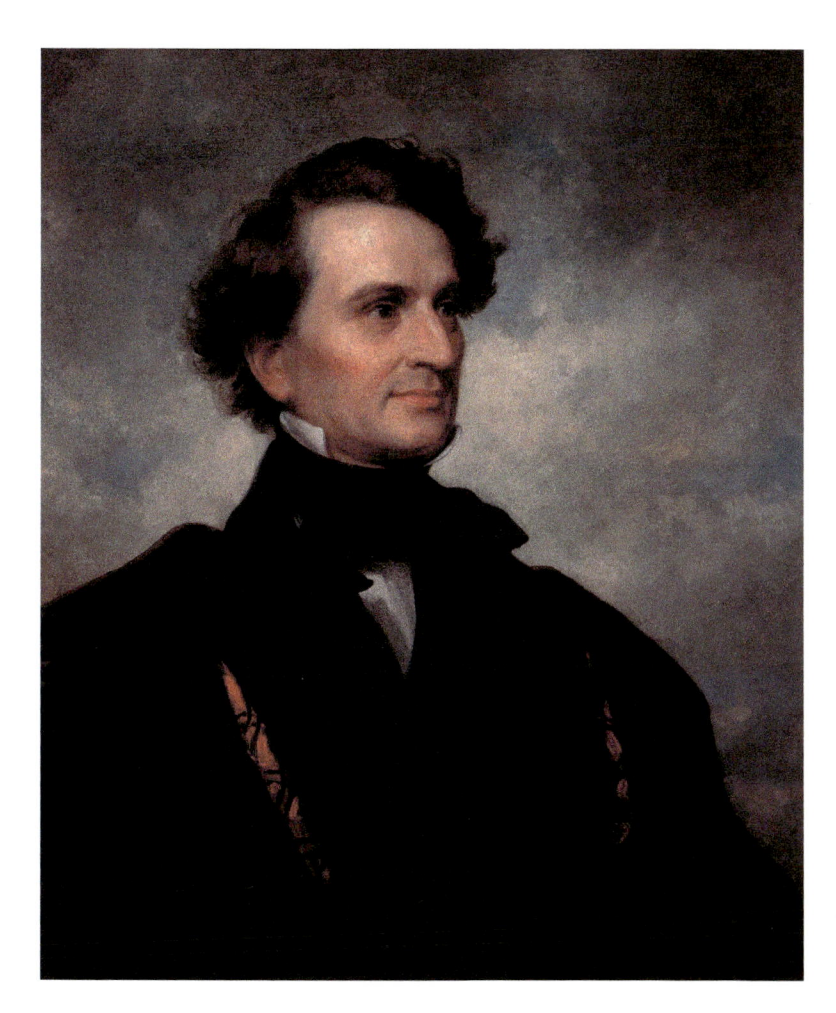

James Dwight Dana

James Dwight Dana, painted by Daniel Huntington in 1858.

The university training of a geologist often includes long hours spent with a hand lens or microscope, studying and describing minerals and learning about their various crystal forms. This is usually undertaken with a key textbook to hand, and for many American geologists, this will have been *Dana's Manual of Mineralogy*. This classic was first published in 1848 and is now in its 23rd edition. A great book remains a great book! Its author, James Dwight Dana, is widely regarded as the foremost American geologist of the second half of the 19th century. He pursued a geological career that diversified far beyond mineralogy; indeed, he was also an outstanding zoologist and explorer with a wide range of accomplishments to his name.

Dana was born in Utica, New York in 1813. From an early age, he enjoyed the outdoors and was fond of collecting rocks, plants and insects. Therefore, it was not surprising that he selected scientific studies at Yale College (now Yale University) in 1830. It was at Yale that he came under the influence of Benjamin Silliman. Silliman was a pioneering chemist, scientific educator and founder of the *American Journal of Science*. He ensured that Dana received an extensive scientific education and served as a mentor to him during his scientific career. In time, he also became Dana's father-in-law.

Dana's sketch of an erupting Vesuvius, published in 1834.

After completing his studies at Yale in 1833, Dana served as a mathematics instructor on a U.S. Navy ship that sailed to the Mediterranean. There, he saw Mount Vesuvius in eruption and sent a report back to Silliman. This report would form the basis for the first paper published by Dana in 1834. In this same year, Dana returned to Yale to take up teaching and research duties. His first objective was mineralogy.

Silliman had an extensive collection of minerals which Dana could supplement with his own specimens, some collected when a child. Using this material, Dana set about developing a new mineral classification based on chemistry and crystallography. His results were published as *System of Mineralogy* when he was just 24. This remarkable work has formed the basis of modern mineralogical studies ever since.

Reminiscent of the opportunity afforded to Charles Darwin to sail on the *Beagle,* in 1838 Dana was offered the opportunity to serve as the geologist and mineralogist of the U.S. Exploring Expedition (1838–1842). For American science, this expedition was without precedent. A fleet of six ships was charged with charting islands in the Pacific, exploring the South American coast and even venturing towards Antarctica. The expedition was substantially funded by the U.S. Navy; and in addition to Dana, there were specialists in botany, vertebrate and invertebrate zoology and philology on board, plus two artists.

For Dana, the voyage was not always comfortable ("*naval servitude*" as he remarked in a letter). The expedition captain had a volatile temperament, and there were the hazards of any voyage in difficult waters. Dana's ship was almost lost in the Straits of Magellan, and aggressive natives made work difficult in Fiji. Nonetheless, he explored the Andes, the atolls of the Pacific and the active volcanoes in Hawaii. Thus, the expedition gave Dana the chance to experience geology on a global scale, which impacted his geological thinking for the rest of his life.

On his return to Yale, Dana's observations were summarised

STAGE I

Diagram showing: Continent with Cold crust, Geosyncline infilling with sediment, Erosion, Ocean, Sea level; Contraction arrows from left and right; downward arrows indicating Cooling & shrinking interior.

STAGE II

Diagram showing: Geosyncline inverted Mts. upheaved, New warping cycle; Contraction continues from left and right; downward arrows.

Dana's Geosynclinal Theory

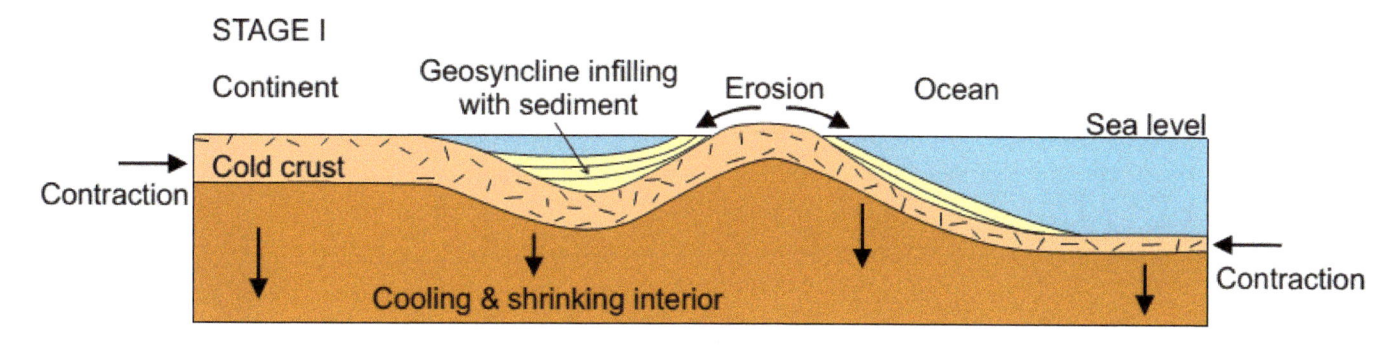

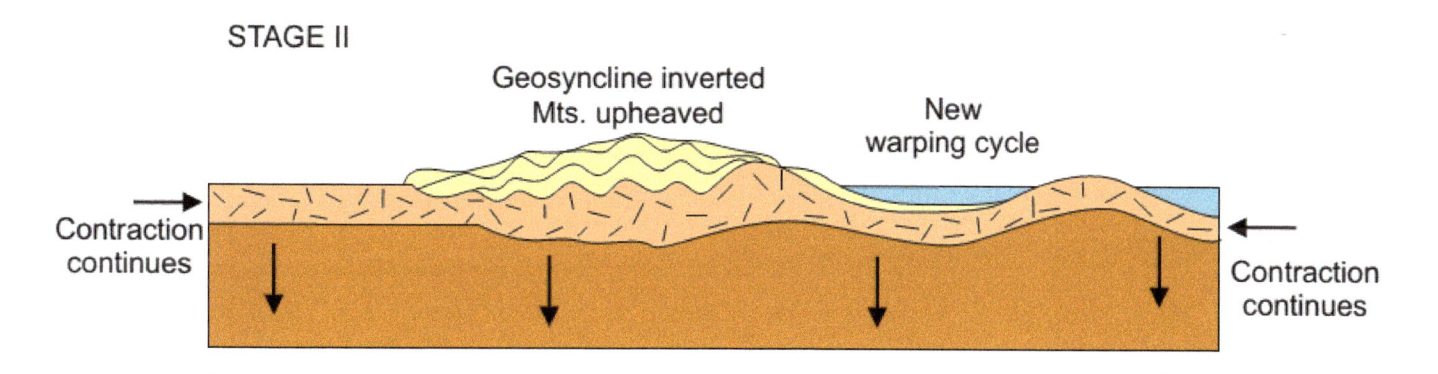

in the expedition's *Report on Geology*, a landmark American geological publication that took thirteen years to complete. In another parallel with Darwin, this report included a discussion of how coral reefs form around volcanic islands. Dana included ideas omitted by Darwin; for example, that atolls may become deeply submerged as a result of subsidence — "guyots" in today's terminology.

In this report and associated papers, Dana produced the first description of many key geological localities. One was Mount Shasta on the California coast. His description of the geology of this area contributed to the California Gold Rush after he had mentioned the likelihood of finding gold deposits in the region.

The contrast between the geology of the Andes and the Pacific Ocean led Dana to consider the continents and ocean basins as separate, permanent and geologically distinct. Arc volcanoes and active mountain belts bound the inactive continental interiors as a result of a cooling, contracting globe. The ocean basins are where volcanic material has vented to the surface and is resisting contraction. Although these ideas are now, in the light of plate tectonics, fundamentally incorrect, they do contain grains of truth that support the notion of a dynamic Earth. For example, Dana's description of linear volcanic island chains in the Pacific led to the hypothesis of Tuzo Wilson that plates pass over hot spots, producing linear island chains in their wake. Plate tectonics has confirmed the contrast in age and structure between continents and ocean basins, and involved the distinctive character of ocean basins, as emphasized by Dana, into the modern global synthesis.

As a result of his thinking on global geology, Dana was engaged in a debate that dominated much of American geoscience in the second half of the 19th century, namely, the processes by which mountains form. The essence of this was the geosynclinal theory, which arose out of a vigorous debate between Dana and the New York geologist, James Hall. Geosynclinal theory can be divided into two main stages. First, large quantities of sediment are deposited within a down-warped part of the crust. Second, these sediments are folded within the mountain building process. Dana described the crustal down-warp area as the "geosynclinal region," and the product of the whole process as the "mountain chain-synclinorium." Dana also emphasized that the trend of the mountain range follows the deepest part of the geosynclinal trough.

There were a handful of suggested hypotheses to support the mechanism of geosynclinal theory. Dana preferred to rely on that of a contracting Earth. Compressional, lateral forces were thought to crinkle up the geosynclinal sediment troughs to produce mountain ranges. James Hall supported what was known as the gravitational sliding hypothesis. This hypothesis relied on isostatic upwarping of geosynclines, paired with the slipping of strata over basement rocks along a flat shear surface. No matter the mechanism, it is important to note that they all embraced purely lateral crustal movement, a concept that would be overtaken by the work of Eduard Suess and other Europeans studying the Alps.

Dana spent his entire academic career at Yale until his death in 1895. During that time, he was a prodigious publisher in the field of geoscience, including seminal textbooks, such as his *Manual of Mineralogy* and his *Manual of Geology* first published in 1863. He also published *Corals and Coral Islands* and *Characteristics of Volcanoes*, the latter completed after a return to Hawaii when he was in his seventies. His legacy was continued by his son, Edward Dana, a noted mineralogist in his own right. Although Dana's name is now mainly associated with his mineralogical studies and textbooks, he was the first American geologist to emphasize the global nature of geology and develop theories to explain the presence of mountains and oceans. These concepts may be outdated now, but they helped set geological science on the path to the plate tectonic paradigm that governs geology today.

REFERENCES

This essay has drawn upon information from the following sources:

Gilman, D.C. 1899. *The Life of James Dwight Dana*. Harper & Brothers Publishers.

Greene, M.T. 1982. *Geology in the Nineteenth Century*. Cornell University Press. 324pp

Natland, J.H. 2003. James Dwight Dana (1813-1895): Mineralogist, Zoologist, Geologist, Explorer. *GSA Today,* February 2003, 20-21.

Oldroyd, D.R. 1996. *Thinking About the Earth*. Athlone.

Oreskes, N. 1999. *The Rejection of Continental Drift*. Oxford University Press. 420pp.

https://publish.illinois.edu/platetectonics/geosynclinal-theory/

Part of the mineral collection of Yale University is now housed in the David Friend Hall of the university's Peabody Museum. The collection was initiated by Dana and his mentor Benjamin Siliman.

Charles Lapworth

Charles Lapworth, c. 1881. Photographer unknown.

The University of Birmingham in the English Midlands houses a small, but highly informative geology museum that was shortlisted for the British Art Fund award, Museum of the Year in 2017. The following quotation greets visitors: "*The problems geology proposes to solve are among the most attractive and the most difficult that can engage the ingenuity of man.*" These inspiring words were written by Charles Lapworth in 1910, and the museum is named after this Great Geologist.

Lapworth received no formal training in geology, but his aptitude in the subject led him to contribute to solving two of the greatest geological controversies of the later part of the 19th century — the so-called 'Highlands Controversy' and the Cambrian–Silurian dispute. Moreover, he demonstrated the utility of biostratigraphy in solving problems of structural geology, pioneered the use of graptolites in stratigraphic calibration and correlation and was a gifted teacher, who did much to encourage women to engage in the science.

Charles Lapworth was born in 1842, in Faringdon, which is now in Oxfordshire. In 1864, after training as a teacher at a college in Culham near Abingdon, he moved to Galashiels in the Scottish Borders to begin a career as a schoolmaster. Troubled by poor health, Lapworth took up geology as an outdoor remedy. He first sought fossils in the local area, but as he became increasingly informed, he began an intensive study of the Southern Uplands of Scotland in 1869.

The Southern Uplands, comprising an apparent great thickness of 'interminable greywackes,' had proven a challenge for the official Geological Survey to map, especially as they undertook rapid reconnaissance traverses to create maps at a typical scale of six inches to one mile. Lapworth took a different approach. He mapped at a minute scale and utilized graptolites as index fossils to characterize different strata. By doing so, he was able to show that the stratigraphy of the region represented many repeated slices, formed by what we would now understand as thrust tectonics.

In 1875, Lapworth moved to St. Andrews to teach at Madras College, but he continued his research. By 1878, the results of his Southern Uplands investigations were published in a

Paleozoic strata at Dob's Linn in southern Scotland. This was a key section that Lapworth used to unravel the stratigraphy and structure of the Southern Uplands.

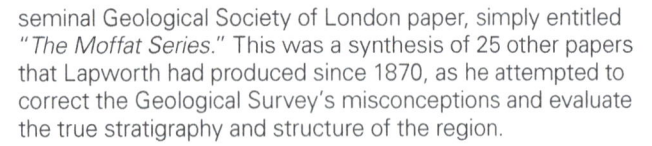

Graptolites from Dob's Linn.

seminal Geological Society of London paper, simply entitled "*The Moffat Series.*" This was a synthesis of 25 other papers that Lapworth had produced since 1870, as he attempted to correct the Geological Survey's misconceptions and evaluate the true stratigraphy and structure of the region.

Central to Lapworth's approach was the detailed documentation of sections where an unambiguous stratigraphic sequence could be located and where characteristic fossils could be defined for each rock unit. Dob's Linn near Moffat was one such locality where he was able to show that a large part of the Lower Paleozoic sequence was condensed into a single exposure about 75 metres thick. This locality is a picturesque gorge with a waterfall and offers very good exposure. Today access is easy, but in Lapworth's time, it was quite remote and he stayed in a small shepherd's cottage. This building had a plaque placed on it in 1931, with the inscription, '*Birkhill Cottage where between 1872 and 1877 Charles Lapworth recognized the value of graptolites as a clue to the geological structure of these hills.*'

In the course of his research at Dob's Linn, Lapworth found that particular graptolite species and assemblages formed excellent markers for specific stratigraphic horizons. This was the key to unravelling the stratigraphy of adjacent areas and to placing the succession in the context of the classic successions of Wales and the Welsh Borderlands. A distinct paleontological break at the base of the rock unit termed the Birkhill Shales contained Llandovery age fossils. Subsequently, this would have significance in defining what was meant by Silurian.

By 1882, Lapworth was able to correlate the rocks of the coastal western Southern Uplands, around Girvan, with the succession to the east near Moffat. He was greatly aided by thirty thousand fossil specimens meticulously collected by a redoubtable local woman, Mrs. Elizabeth Gray. The correlation was difficult because the lithologies and thicknesses were very different. Around Girvan, the succession is several hundred metres thick and contains many thick sandstones beds; near Moffat, the same beds are represented by tens of metres of shales. Lapworth recognized that this was because of paleogeography. The sediments to the west were more proximal to a sediment source, whilst those to the east were more distal and represented condensed deposition. In modern terminology, Lapworth was recognizing the margin of the ancient Iapetus Ocean.

Lapworth's conclusions were, initially, an embarrassment to the Geological Survey and their director, Sir Archibald Geikie, who had promoted much simpler explanations of the succession. However, talented field geologists, such as John Horne, saw the value of Lapworth's methods and conclusions, and employed them in the Survey's remapping of the area.

The conclusions of Lapworth's research in the Southern Uplands had a significant bearing on an acrimonious debate that had marred Paleozoic research efforts from the mid-19th century. Once friends and collaborators, two titans of Victorian geology, Sir Roderick Murchison and Adam Sedgwick, disputed what was meant by terms they had each introduced: Silurian by

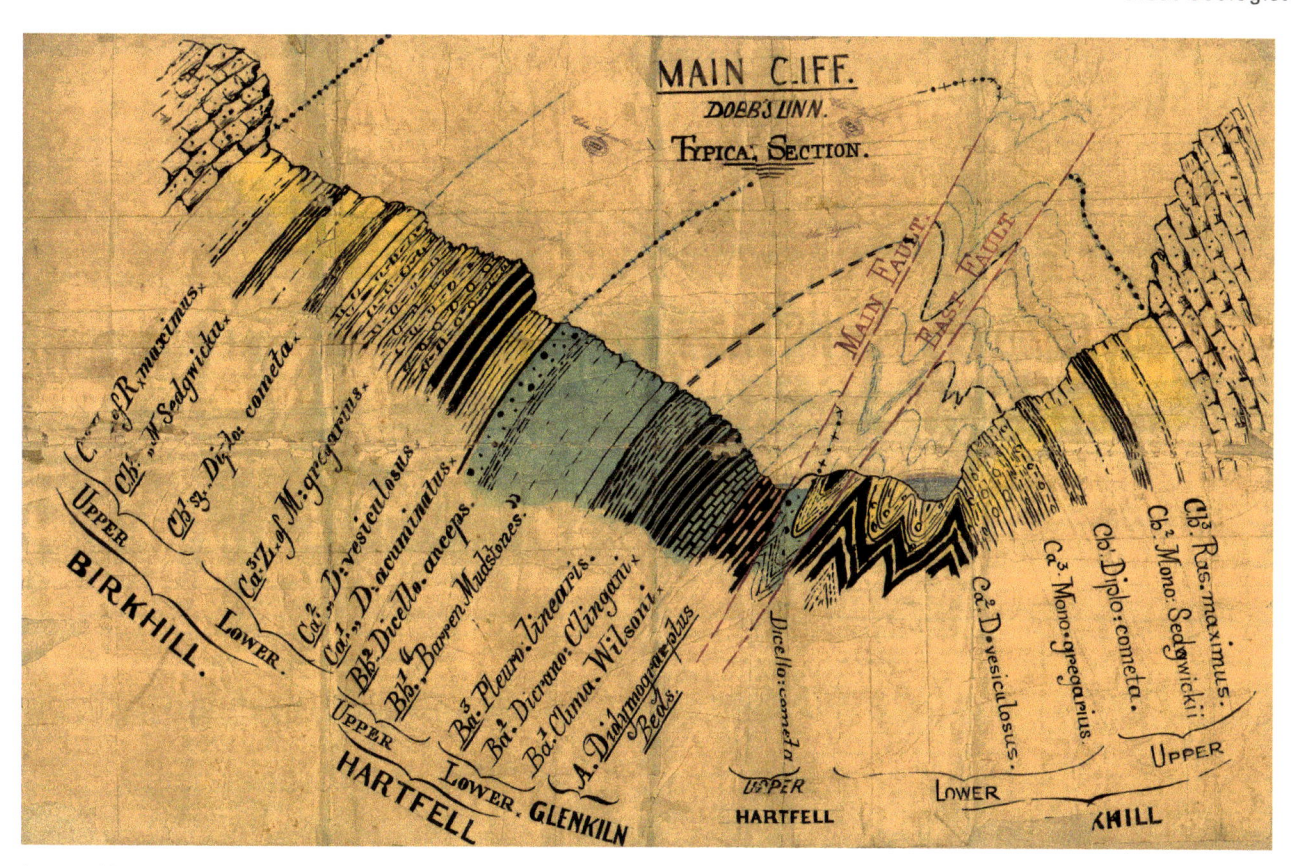

Lapworth's original hand drawn and coloured interpretation of the classic Dob's Linn outcrop, key to his understanding of Paleozoic stratigraphy and structure in the Southern Uplands. Image provided and used with the permission of the Lapworth Museum of Geology.

Murchison and Cambrian by Sedgwick. Put simply, Murchison wished to extend Silurian downwards to the base of all strata containing fossil remains, and Sedgwick wished to extend Cambrian upwards to the base of an unconformity seen in the Welsh Borderlands (the base of Murchison's Upper Silurian). Rock units termed the Llandeilo and Caradoc series were claimed by Murchison for the Silurian and by Sedgwick for the Cambrian.

From his knowledge of fossil faunas, Lapworth took a straightforward approach to resolving this problem. The Lower Palaeozoic could simply be divided into three main epochs —Cambrian and Silurian, with a new term, Ordovician, for the strata lying between them. Each was readily definable in terms of biostratigraphy. This approach was outlined in 1879 (after the deaths of both Sedgwick and Murchison). It found favour with many academic geologists, but was resisted in the Geological Survey, which was led by Murchison's acolyte, Geikie. After Geikie's death in 1901, the Survey began to use the term Ordovician officially, although it took until the meeting of the International Geological Congress in Copenhagen in 1960 for it to be fully sanctioned by the international geological community! Academic stratigraphic geology is not always known for rapid decision making. Incidentally, the Global Stratotype Section and Point (GSSP) for the base of the Silurian system was placed within the Dob's Linn section in 1985.

By 1881, Lapworth's standing in the geological community was such that he was appointed Professor of Geology and Mineralogy at Mason's College in Birmingham, the forerunner to the University of Birmingham, which was founded in 1900. He served the university with great distinction until his retirement in 1913, building a department that became one

Mrs Elizabeth Grey, 1922. Her extensive fossil collections proved invaluable to Lapworth. Provided by and used with permission of The Natural History Museum.

of the leading centres for geological teaching and research in Britain.

Notwithstanding his location in the Midlands, Lapworth continued to focus his research on Scotland, now turning his attention to the Northwest Highlands. This region had been the subject of a long-standing controversy, with Murchison once again on one side of the debate. Murchison considered the succession there to be effectively continuous, despite metamorphic schists overlying unmetamorphosed limestones, shales and sandstones. He claimed much of this succession for his Silurian system. Others, such as James Nicoll of Aberdeen University, considered the contact at the base of the schist to be tectonic. Lapworth decided to see if his detailed mapping methods employed in the Southern Uplands could be used to shed light on this problem.

Lapworth carried out two intense field seasons in 1882 and 1883, in the area in the far north-west of Scotland, around Durness and Loch Eriboll. In his mind were some of the ideas on mountain building, including what would become known as thrust tectonics, being promoted in continental Europe by geologists, such as Escher, Heim, Suess and Brøgger. He was familiar with reverse faulting from his observations in the Southern Uplands (such faults are present at Dob's Linn, for example).

As was his practice, Lapworth began mapping around Durness at a very detailed scale — twenty-five inches to the mile in some cases — before moving to a shepherd's cottage at Heilam near Loch Eriboll. There, he found an outcrop at An t-Sròn that could act as a stratotype for the region (in a similar manner to the function of Dob's Linn in the Southern Uplands). Unlike the Southern Uplands, the succession in the Northwest Highlands is poorly fossiliferous, so Lapworth was forced to focus on detailed lithology to build up his understanding of the stratigraphy.

Lapworth soon recognized that the contact between the metamorphosed schists and the underlying sediments was tectonic. Indeed, the contact was often characterized by a specific new rock type formed by actions of shearing — mylonite. In a letter to his friend and colleague, Thomas Bonney, he describes such rocks as follows: "*Conceive a vast rolling and crushing mill of irresistible power, but of locally varying intensity, acting not parallel with the bedding but obliquely thereto; and you can follow the several stages in imagination for yourself. Undulation, corrugation, foliation and schistose structure – slaty cleavage are all the effects of one and the same cause…..Shale, limestone, quartzite, granite and the most intractable gneiss crumple up like putty in the terrible grip of this earth-engine – and are all finally flattened out into thin sheets of uniform lamination and texture.*"

By 1883, Lapworth was able to publish a preliminary account of his interpretation, noting the structure to be dominated by "*gigantic overfolds*" and "*gliding-planes, along which the rocks have yielded to the irresistible pressure of the lateral Earth-creep during the process of mountain-making.*" This was a first description of large-scale thrust tectonics in British geology. Lapworth called it, "*The Secret of the Highlands.*"

Towards the end of his 1883 field season, Lapworth suffered what appears to have been a nervous breakdown. The popular story is that he imagined the great Moine Nappe grating over his body as he lay tossing in his bed at night. Perhaps, he was concerned about another intellectual fight with Geikie, who, as ever, supported the simplistic Murchison view of the succession (i.e. all conformable). How true this is uncertain, but after 1883, he left research in the region to others. Having clarified the large-scale structure, the details would be fully realized by the Geological Survey team led by the legendary team, Peach and Horne.

The later part of Lapworth's career focused on the Paleozoic geology of the Midlands and Welsh Borderlands, where his application of detailed biostratigraphic zonations extended downwards into Cambrian strata.

Lapworth's research on the graptolites, which began during his work in the Southern Uplands of Scotland, continued throughout his career, and he became world renowned as the

Charles Lapworth (centre) leading a field excursion to the Severn Valley in the 1880s. Image provided and used with the permission of the Lapworth Museum of Geology.

foremost authority on this extremely important fossil group. He published numerous important scientific papers on this group of fossils and assisted geologists worldwide with the identification, dating and interpretation of graptolite faunas.

Lapworth was regarded as an excellent teacher and mentor, notably encouraging women to take up geological research. In particular, his student and assistant, Ethel Wood (later, Dame Ethel Shakespear), and her collaborator, Gertrude Elles, carried on his graptolite research, authoring the seminal *Monograph of British Graptolites* (1910) under his supervision. Much of his spare time was given to leading field excursions in the West Midlands, not only for formal students, but also for natural history societies and other amateur groups. Once an 'amateur' himself, Lapworth always promoted the value of what those with a genuine interest in geology could achieve, whether or not they had been formally trained.

Lapworth received many awards for his work and contributions to geology. In June 1888, he was elected a Fellow of the Royal Society and, in 1891, he was awarded their Royal Medal. In 1899, he received the highest award of the Geological Society of London, the Wollaston Medal, in recognition of his outstanding work in the Southern Uplands and Northwest Highlands of Scotland.

ACKNOWLEDGEMENTS

The author would like to thank Jon Clatworthy, Director, Lapworth Museum of Geology for his kind help in sourcing some of the images used in this biography.

REFERENCES

This essay has drawn upon information from the following sources:

Fortey, R.A. 1993. Charles Lapworth and the biostratigraphic paradigm. *Journal of the Geological Society, London*, 150, 209-218.

Hamilton, B. 2001. Charles Lapworth's "The Moffat Series", 1878. *Episodes*, 24, 194-200.

Oldroyd, D.R. & Hamilton, B.M. 2002. Themes in the early history of Scottish geology. In: Trewin, N.H. (ed.) *The Geology of Scotland*. The Geological Society, London, 27-44.

Oldroyd, D.R. 1990. *The Highlands Controversy.* The University of Chicago Press. 438pp.

Rider, M. 2005. *Hutton's Arse*. Rider-French Consulting Ltd. 214pp.

https://www.birmingham.ac.uk/facilities/lapworth-museum/about/lapworth.aspx

Henry Clifton Sorby

Henry Clifton Sorby photographed in the 1860s. Photographer unknown. Image courtesy of the University of Sheffield.

Many of the standard procedures employed in modern geoscience we now take for granted. One of these is the study of thin-sections, that is to say, small slices of rock ground to a thickness of 30 microns so that light can pass through allowing mineral composition to be determined with the use of a petrographic microscope. Another is the study of sedimentary structures to assist in the determination of the depositional environment of a sedimentary rock. But who first thought of these techniques? These and many other geological and scientific innovations came from the work of a brilliant 'gentleman scientist' of the Victorian age – Henry Clifton Sorby.

The advancement of science in the 19th century was in no small part due to the work of gentlemen of independent means who were able to set up offices and research laboratories at home and develop a network of correspondents (long before the advent of Google Scholar and email). They harnessed their curiosity for the natural world and, without being tied to any academic institution, researched and published prolifically. Some were gifted polymaths who happily turned their focus of interest from one branch of science to another as their curiosity led them. One such person was Sorby whose achievements in geology were substantial, but who also published with distinction on a wide range of subjects including metallurgy, biology and archaeology.

Sorby was a genuine innovator in geology. It can be said that he founded petrography through the microscopic study of thin-sections and sedimentology through the study of sedimentary structures. He also pioneered the use of fluid inclusions to understand the formation and burial history of rocks and minerals. Outside of geology, he founded the microscopic study of metals (metallography) that revolutionized the production of steel, and he created a spectroscopic microscope to analyse pigments and staining (which also had applications in forensic science). Most of this work was carried out at his home in Sheffield where he was based for his entire career.

The city of Sheffield in England has long been famous for its metal work, and the Sorby family owned a well-established tool manufacture business there. Henry Clifton Sorby was born in 1826. He might have been expected to run the family business in due course, but during his schooling he had shown an aptitude for science and mathematics. His father died unexpectedly in 1847 and so at the age of 21 he inherited the family fortune. Sorby used the monies he had been left to create a scientific laboratory at the home he shared with his mother. He was devoted to his mother for the rest of her life, and she either accompanied him on his travels or he never left her for more than four days at a time. In return, she encouraged him in his research.

During the early part of his career, the primary outlets for his endeavours were the meetings and reports of the Sheffield Literary and Philosophical Society. His first paper published in the reports of this society was on agricultural chemistry in 1846, but a year later geology had become his main scientific focus. In 1847 he published on fluvial geomorphology in the Sheffield region. In 1848 he chanced to have a meeting during a train journey that was to alter profoundly the course of his research. The person he met was William Williamson, a physician and jeweller (and subsequently an expert in stratigraphic paleontology, paleobotany and foraminifera), who made thin-sections of petrified wood, teeth and bones. Sorby soon learned the preparation technique at Williamson's home and immediately started preparation of thin-sections of ordinary rocks. Although the microscopic study of thin-sections was not new (William Nicol had introduced a polarizing microscope several years previously), the technique had largely been used to study the structure of bones and fossils and never used systematically to study rocks and minerals. Thus Sorby founded microscopic petrography by using thin-sections of rock ground to a thousandth of an inch to study their composition and origin.

By 1849 Sorby was able to present the earliest results of his petrographic research to his local Sheffield society and in 1851 published his first paper on the subject, discussing some sedimentary rocks from the Yorkshire coast, where he distinguished quartz, chalcedony and calcite apart by use of polarized light on a rotating stage. His early efforts were not always well received by the scientific community at large. He later said "*In those early days people laughed at me. They quoted Saussure who said it was not a proper thing to examine mountains with a microscope, and ridiculed my actions in every way. Most luckily I took no notice of them*". These attitudes were to change when Sorby tackled the problematic origin of slaty cleavage in metamorphosed sediments that can cut across primary sedimentary fabric. From the microscopic study of slates, he was able to determine that the cleavage originates from the reorientation of mica under anisotropic pressure as the original sediment undergoes deep burial. This research (published in 1853 and 1856) brought him to the attention of the scientific establishment, and he was duly created a Fellow of the Royal Society when only 31 years old.

Thin-section of an Oligocene limestone from north-western Turkey

During the next few years Sorby published a series of brilliant papers on the petrography of sandstones and limestones, but his interests also included igneous and metamorphic rocks (for example, he was the first to describe the deformation of ooids and crinoid ossicles in lightly metamorphosed carbonates). He determined different types of detrital quartz and how this could be linked to sediment provenance. In his papers on carbonate rocks he described the conversion of aragonite to calcite, dolomitization, different phases of cementation and recognized that the coccoliths that form chalk are of organic origin. He described many of the constituent grains of carbonate rocks including ooids and a variety of bioclasts.

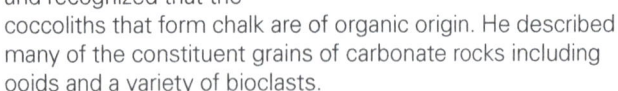

During a visit to Germany in 1860, Sorby met a young German geologist named Ferdinand Zirkel who was inspired by the new science of petrography and applied it systematically to igneous rocks – the results are summarized in the 1866 publication *Lehrbuch der Petrographie*. By this time Sorby's interests were already moving on, first to the petrography of meteorites and then to man-made metals.

Sorby recognized that metals had a crystalline structure and the composition of steel could be determined by microscopic examination once it had been etched by acid. His understanding of the composition of steel and what gave it its strength revolutionized its manufacture. He later remarked "*In those early days, if a railway accident had occurred and I had suggested that the company should take up a rail and have it examined under the microscope, I should have been looked upon as a fit man to send to an asylum. But that is what is now being done.*"

Parallel to his petrographic studies, Sorby was engaged in a number of other lines of geologic research. Perhaps the most important of these was the recognition of sedimentary structures as indicators of past deposition processes, thereby founding the science of sedimentology. He first noticed current-formed structures whilst sheltering from the rain in a quarry in 1847. This was at the time that he was researching fluvial geomorphology for his first publication, and he was struck by the similarities between the structures he observed in the rock and those he had observed forming in modern sediments. Over the next few years, he systematically observed and described sedimentary structures at outcrop and conducted flume tank experiments at home to understand their genesis. A series of publications emerged - for example "*On the structures produced by the current present during the deposition of stratified rock*" published in 1859, but a full summary of this line of endeavour was not published until his monumental final paper in 1908, "*On the application of quantitative methods to the study of the structure and history of rocks.*". This masterly synthesis of the hydrodynamic interpretation of sedimentary structures reviewed their relationship with current velocity, the angles of repose of different sediments and the settling velocity of grains in water. It also described soft sediment deformation, the measurement of porosity and the exact determination of structural deformation.

The use of fluid inclusions in minerals to understand the temperature of crystallization is now a well-established technique in a variety of geological studies. Sorby was, once again, a pioneer in developing this technique, publishing a key paper "*On the microscopical structure of crystals, indicating the origin of minerals and rocks*" in 1858.

In 1878, Sorby purchased a yacht, *The Glimpse*. He had it fitted out as a floating laboratory and, over a series of twenty-five summers, proceeded to survey the eastern coast of England, carrying out studies on coastal sediments and marine fauna. His intellectual curiosity seems to have been boundless. In 1892, he walked 1,200 miles in four months (aged 66) studying and describing ancient buildings for his archaeological research.

Thin-section of a calcareous cemented sandstone. Several different types of detrital quartz grains are present which could be related to different sediment source regions, a notion that Sorby pioneered.

Climbing ripple lamination indicative of high sediment flux under strong flow conditions.

Sorby dedicated his entire adult life to science and at the time of his death in 1908, he had published around 250 papers on a diverse range of subjects. A model of what a creative individual can accomplish working alone, he was honoured by The Royal Society, The Geological Society and countless other scientific bodies. Both the International Association of Sedimentologists and the International Metallographic Society have annual awards for outstanding achievement in his name. He promoted the idea of a university in Sheffield and lived long enough to see it founded in 1905. Upon his death, he bequeathed much of his fortune to this new academic institute. As the title of one his biographies states, he was indeed *"A Very Scientific Gentleman."*

REFERENCES

This essay has drawn upon information from the following sources:

Allen, J.R.L. 1993. Sedimentary structures: Sorby and the last decade. *Journal of the Geological Society, London*, 150, 417-425.

Eyles, J.M. 1951. William Nicol and Henry Clifton Sorby: Two Centenaries. *Nature,* 168, 98-99.

Folk, R.L. 1965. Henry Clifton Sorby (1826-1908), the Founder of Petrography. *Journal of Geological Education*, 13(2), 43-47.

Higham, N. 1963. *A Very Scientific Gentleman*. Macmillian, 160pp.

Smith, C.S. 1960. *A History of Metallography.* University of Chicago Press, 291pp.

http://www.sorby.org.uk/about-us/henry-clifton-sorby/

https://www.ypsyork.org/resources/yorkshire-scientists-and-innovators/henry-clifton-sorby/

https://www.theguardian.com/education/2007/feb/08/highereducation.comment

Thrust-related folding in the Alps.

Eduard Suess

It is arguable that Eduard Suess is one of the greatest geologists that ever lived, yet many geologists who are active today are unaware of him, although they routinely use terms first coined by him. Eustasy, Gondwanaland, Tethys Ocean, foreland, listric fault, horst, graben, batholith, island arc and many other terms in common use today originated from his pen. He published many significant works but none greater than *Das Antlitz der Erde* (The Face of the Earth) which in three main volumes, numbering 2788 pages, and developed over almost 30 years, he attempted to review global geology, especially tectonics and stratigraphy in their broadest sense. Remarkably, at the same time as researching this monumental work, he enjoyed an important political career and fostered scientific thinking in the Germanic late 19th century world through his leadership of the Austrian Academy of Sciences. Not surprisingly, he is regarded as one of the greatest scientists and polymaths of the German-speaking world.

Born in London in 1831 (where his parents had briefly moved to from Vienna) he was three when he travelled first to Prague and then to Vienna where he spent the rest of his life. Self-taught in geology and paleontology, in 1852 he was appointed as an assistant at the Hofmineralenkabinett in Vienna, a practical school of geology and mineralogy and soon began to publish scientific papers, the first being papers on graptolites, brachiopods and ammonites. This led to him being appointed Professor of Paleontology at the University of Vienna in 1857 and then, as his interests in geology broadened, to becoming Professor of Geology in 1862. He remained on the staff of the University till his retirement in 1901, a hugely popular figure with both students and colleagues.

As might be deduced from the title of his *magnum opus*, the greater part of his scientific career was devoted to working out the evolution of the features of the earth's surface, and in particular mountain-building. Even so, his

Eduard Suess c. 1869, lithograph by Josef Kriehuber.

The Greek Titaness Tethys, for whom Suess named the Tethys Ocean. Fourth-century mosaic from Philipopolis. Shahba Museum, Syria.

interests were incredibly broad and included what today we would call ecosystems, earth system science and urban geology (his first major work was in the geology and soils of Vienna). He published well-received papers on the geology and economics of gold and silver.

Notwithstanding these broad interests, tectonics was a key focus. His fieldwork often took him to the eastern Alps and Carpathians (he was an accomplished mountaineer) and there he started to develop ideas to challenge the then popular notion that vertical movements were largely responsible for their creation (i.e. they represent folds around intrusive masses). Instead, he emphasised the importance of horizontal movements, although in a pre-plate tectonic era he ascribed these to the contraction of the Earth through cooling which was a prevalent concept of global geology in the second half of the 19th century. He first addressed these issues in a small book published in 1875, *Die Entstehung der Alpen* (The Origin of the Alps), before expanding them in *Das Antlitz*. He wrote (in translation): "...*the entire surface of the Earth really moves slowly and inhomogenously... the so-called mountain massifs thereby move more slowly than the regions lying between them that pile up and build mountain chains. In Central Europe they create regular folds on their polar side and tears on the equatorial side*". In effect he is describing simultaneous shortening and extension resulting from differing depths of décollement, a tectonic process well established today.

Fluent in several languages, Suess was one of the great pioneers of geological synthesis, absorbing a vast amount of international literature. This enabled him, for example, to identify the paleontological similarities between Africa, Australia, Madagascar, India and South America which led to the concept of Gondwanaland (although joined by land bridges, not by moving continents). He recognised that Gondwanaland had been framed by the Tethys Ocean; an ocean he understood had been reduced to the modern day Mediterranean Sea. Such ideas must have proved useful to Alfred Wegener as he developed his theory of continental drift, the forerunner of modern plate tectonics.

At the time Suess was carrying out his initial research the geoscientific community was still divided between "catastrophists" following Cuvier and d'Orbigny that viewed the history of the Earth as a series of discreet episodes, partitioned by global events; and "uniformitarianists", following Lyell, that preferred a steady-state Earth, where gradual, local processes dominated. Suess found his initial uniformitarianist stance challenged by his own observations. Evaluating the Neogene stratigraphy of the Vienna Basin, which had been assumed to be distinctively local, forming in response to isolation by uplift of the Alps and Carpathian Mountains, he noted that the stratigraphy matched that of the Black Sea, the Caucasus and even as far east as the Aral Sea. This could not, therefore, simply be the product of local tectonics; it required fundamental changes in sea-level. Exploring this concept further, he noted that certain periods of geological time seemed to be associated with transgression, others with regression. In *Die Entstehung der Alpen* he noted the importance of transgression during the Cenomanian, for example. Suess explained these transgressions and regressions in the context of an episodically contracting Earth that would create subsidence that would in turn increase the capacity of the ocean basins and withdraw the ocean from the continental edges (i.e. a regression). When the space created was filled with sediments the ocean would be displaced onto the edge of the continents (i.e. a transgression). We know now that this mechanism is of course wrong, but it does not invalidate the primary observations that Suess described as resulting from "eustatic movements" ("*eustatische Bewegungen*"). Indeed during his lifetime Suess noted that the mechanisms to explain eustasy remained highly uncertain. This is still the case today.

Suess cared a great deal for his home city of Vienna and in 1863 he joined the municipal council. One of the projects he pioneered was the improvement of the water supply to the city, initiating the construction of an aqueduct 110 km long from the Alps. This led to a remarkable reduction in typhoid and cholera cases in the city. During this time he also directed the regulation of the Danube to prevent flooding in the city, a project that proved successful even well into the 20th century. In 1873 he joined the Austrian Reichstag (parliament) as a deputy where he served till 1897. Combining the career of academic geologist with politician required a strong appetite for hard work and Suess clearly possesed this. Waking at half past six, he walked to the university where he held lectures from eight till nine. He then worked in the morning and used the afternoon for brief rest and thinking time before working in the evening and into the night preparing papers and lectures. He married in 1855 and had seven children. The support of his family was obviously crucial to his success.

His style was not to present a short compelling theory but rather to place his interpretations within long and detailed regional descriptions, inviting the reader to draw their own conclusions. This led to his work being considered more of an encyclopaedia to be dipped into rather than a theory to be evaluated. This may in part explain why he is not as well-known as some of his contemporaries, especially outside of German-speaking nations. "*What I offer you is little more than a number of questions; but questions are the buds on the tree of knowledge*" he wrote. This modest self-assessment underplays the contribution of an incredibly productive man whose ideas still remain crucial today.

REFERENCES

This essay has drawn upon information from the following sources:

Gohau, G. 1990. *A History of Geology.* Rutgers University Press. 259pp.

Greene, M.T. 1982. *Geology in the Nineteenth Century.* Cornell University Press. 324pp.

Hallam, A. 1992. Eduard Suess and European Thought on Phanerozoic Eustasy. In: Dott, R.H. Jr. (ed.) *Eustasy: The Historical Ups and Downs of a Major Geological Concept.* Geological Society of America Memoir, 180, 25-30.

Hofmann, T., Blöschl, G., Lammerhuber, L., Piller, W.E. & Şengör, A.M.C. 2014. *The Face of the Earth: The Legacy of Eduard Suess.* Edition Lammerhuber, 105pp.

Oldroyd, D.R. 1996. *Thinking About the Earth.* The Athlone Press, 410pp.

Şengör, A.M.C. 2014. Eduard Suess and global tectonics: an illustrated guide. *Austrian Journal of Earth Sciences*, 107, 6-82.

Wagreich, M. & Neubauer, F. 2014. The geological thinking of Eduard Suess (1831-1914) between basic research and application: an introduction. *Austrian Journal of Earth Sciences*, 107, 4-5.

T. C. Chamberlin

"Is geology an art or a science?" I was posed this question during my recruitment interview with the Chief Geologist of the supermajor that I worked for in the early years of my career. The question may seem odd to many. Geology is obviously a science, isn't it? I presumed my interviewer was hinting at the need for creativity in interpretation of geologic data. But *how* geology functions as a science is a question that has intrigued many including, arguably the greatest American geologist of his age, Thomas Chrowder Chamberlin.

Karl Popper suggests that the scientific method is based on the premise that a simple, unbiased observation promotes a hypothesis to explain it. This hypothesis can be tested by experimentation and further observation which can falsify it, with the emphasis on falsification. In practice, much of geological science does not follow this process. Rather a hypothesis gathers support from observations — geoscientists view data in the light of the hypothesis they particularly support, often linked to their experience. Consider, for example, the competing hypotheses of a contracting Earth and plate tectonics to explain mountain formation as debated in the 20th century. There may always be an element of bias as hypotheses gain support or are refuted. Indeed, some view this as a good thing. *"No progress without prejudice"* remarked A.F. Buddington, the great Princeton petrologist, meaning that championing a hypothesis helps gather important data and spurs our science forward.

In 1890, Chamberlin published on *"The Method of Multiple Working Hypotheses"* — a landmark and still-cited paper. He encouraged geologists (indeed all scientists) to carry multiple explanations for their observations in their minds and weigh then equally until further observations promoted one above all others. So, to explain a sedimentary breccia deposit observed in the field, mass transport, faulting or dissolution of matrix sediment, are all equally plausible explanations until

T.C. Chamberlin from an 1897 photograph. Photographer unknown.

definitive evidence is observed. We should, he argued, avoid bias from the guidance of a "ruling theory". Such an innocent approach is probably implausible in practice. We interpret rocks through the lens of our experience. Nonetheless, it is obviously good scientific practice to keep an open mind.

Chamberlin was very much more than a scientific theorist. He was an exceptional field geologist who recognised the record of glaciation in the American landscape and Pleistocene sedimentary record. Somewhat ahead of his time, he considered the role of atmospheric gases in climate change and their sequestration in the rock record. He also developed a theory for the formation of the Earth linked to the tectonic processes responsible for mountain building, sea-level change and the subdivision of the geological record! In his own words, he sought to explain *"the very soul of geologic history."* He was also an excellent teacher, writer and academic administrator and began publication of the still-influential *Journal of Geology.*

Chamberlin was, literally, *"born on a moraine"* in southeastern Illinois in 1843. Soon after, his father, a Methodist circuit minister and farmer, moved the family near to Beloit in southern Wisconsin. Chamberlin excelled as a scholar at the local college and initially became a teacher, and later principal, of a local high school. By 1873, he was back at Beloit College as professor of geology, zoology and botany.

1873 also marked the beginning of his serious research endeavours, beginning a comprehensive geological survey of Wisconsin and in particular its glacial geology. By 1876, he was Chief Geologist of the Wisconsin Geological Survey and within six years had completed a large four-volume treatise of the geology of the state. This brought him national attention and he was appointed Head of the glacial division of the National Survey in 1881. His research led him to be the first to demonstrate that there had been multiple Pleistocene glaciations in North America. Using features such as moraines, drumlins and eskers, he was able to map the limits of the last two glacial advances.

His full-time activities with the U.S. Geological Survey proved to be short-lived. His organisational and administrative skills led him to being invited to be President of the University of Wisconsin at Madison. This would prove to be a break from his research, but he undertook the role from 1887 to 1892, greatly reforming the university for the better.

In 1892, the opportunity arose for him to return to geology full time with an offer to organise a department of geology at the new University of Chicago. He was to remain there until his retirement in 1918, creating and leading a distinguished faculty and research programme. From there he launched *Journal of Geology*, partly as a vehicle for the prolific outpouring of his scientific ideas.

His range of activities was immense, although a common theme was the origin of the Earth and how this influenced Earth's history, structure and processes. At the heart of this was his planetisimal theory for the origin of the planets, developed with the Chicago-based astronomer Forest Moulton. In this theory, immensely hot matter expelled from the sun and dragged out by the gravitational attraction of a passing star accreted to form cold objects (planetisimals). They acquired an orbital motion around the sun and periodically collided to accrete even further, eventually to form the planets that exist today. Once formed, the Earth was subjected to continuing, but episodic, radial gravitational contraction. For Chamberlin, these contractional events caused mountain building due to differential vertical movements of radial Earth segments. Oceanic subsidence caused less dense continental blocks to move upwards in accordance with isostatic principles. Chamberlin's Earth was solid to its core, so there was no fluid interior on which blocks of crust could founder.

Episodic tectonic events provided a control on global sea-level and continental erosion, and in turn explained the natural events that led to the biostratigraphically-based subdivisions of geologic time (regression leading to extinction; transgression leading to new fossil species occupying newly created niches). There are echoes of the catastrophism of

Chamberlin's 1894 map of Pleistocene ice extent in North America based on the mapping of moraines and other glacial landforms.

Georges Cuvier and Alcide d'Orbigny and an anticipation of aspects of modern sequence stratigraphy.

In 1899, Chamberlin challenged Lord Kelvin's famous calculation of the age of the Earth as being only 20–30 million years, based on the time of cooling from a molten origin. With his cold, planetisimal origin of the Earth in mind, he argued that some unknown source of heat energy within the Earth would alter Kelvin's calculations substantially. This was a prescient anticipation of heat from radioactive decay that in the coming decades would establish the immense age of the Earth.

At time of his death in 1928, he had contributed to around 250 publications including the textbook *Geology* (co-authored with Rollin Salisbury), which was the *de facto* standard English-language geological textbook of the early 20th century until the publication of *Principles of Physical Geology* by Arthur Holmes in 1944. He was the recipient of the first Penrose Medal of both the Society of Economic Geologists (1924) and the Geological Society of America (1927).

Chamberlin had a great physical and intellectual presence and the ability to argue eloquently. He could be damning in his criticism of his scientific opponents, which takes us back to his "method of multiple hypotheses" — did he practice what he preached? Perhaps not, he was certainly dogmatic in his defence of his planetisimal theory. As Robert Dott, Jr. remarked in a biographic article *"the great T.C. Chamberlin was human after all."*

REFERENCES

This essay has drawn upon information from the following sources:

Chamberlin, R.T. 1932. Thomas Chrowder Chamberlin 1843-1928. *National Academy Biographical Memoirs*, 15, 307-407.

Dott, R.H. 2006. Rock Stars: Thomas Chrowder Chamberlin (1843-1928). *GSA Today*, October 2006, 30-31.

Dott, R.H. Jr. 1992. T.C. Chamberlin's hypothesis of diastrophic control of worldwide changes of sea-level: a precursor of sequence stratigraphy. In: Dott, R.H. Jr. (ed.) *Eustasy: The Historical Ups and Downs of a Major Geological Concept.* Geological Society of America Memoir, 180, 31-42.

Greene, M.T. 1982. *Geology in the Nineteenth Century.* Cornell University Press. 324pp.

Hallam, A. 1983. *Great Geological Controversies.* Oxford University Press, 244pp.

Le Grand, H.E. 1988. *Drifting Continents and Shifting Theories.* Cambridge University Press. 313pp.

Oldroyd, D.R. 1996. *Thinking About the Earth.* The Athlone Press, 410pp.

Oreskes, N. 1999. *The Rejection of Continental Drift.* Oxford University Press. 420pp.

Powell, J.L. 2015. *Four Revolutions in the Earth Sciences.* Columbia University Press, 367pp.

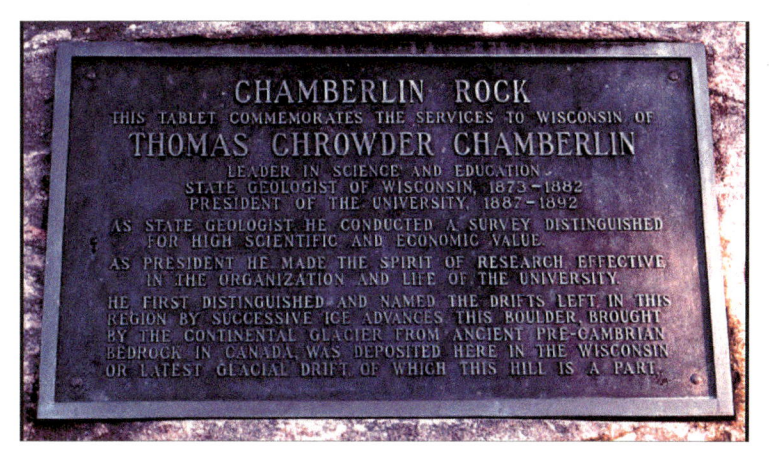

Memorial plaque commemorating Chamberlin at the University of Wisconsin. Appropriately, it is place on a large glacial erratic boulder.

Alexander Karpinsky

The first Sunday of April is celebrated as Geologists Day in Russia and other countries of the former Soviet Union. This commemorative day was introduced to acknowledge the tremendous contribution geologists make to society, both in terms of developing scientific understanding and, more practically, through the applied aspects of the subject. Its timing in spring reflects when geologists start to plan and begin their field work. As might be expected within a huge country with diverse and outstanding geology, Russian geologists have made significant and telling contributions to the science. Foremost amongst these is Alexander Karpinsky, who pioneered the geological mapping of Russia, locating its mineral resources and contributing across an outstanding breadth of geoscience – from paleontology and mineralogy to global tectonics.

Alexander Petrovich Karpinsky was born in 1847, in the town of Turyinskiye Rudniki (now Krasnoturyinsk within Sverdlovsk Oblast) located in the Ural Mountains. This was a copper mining settlement on the Turya River, and Karpinsky's family were mining engineers. Copper had been mined in this region since the Bronze Age. Not surprisingly, Karpinsky was encouraged to follow in the family tradition, and when only 11 years old, he began studies at the Saint Petersburg Mining Institute, eventually graduating from the Mineralogical Institute in the same city in 1866. He then returned home to work as a mining engineer in the Urals.

Karpinsky's work as a practical mining geologist was brief. His intellectual talent had been noted in Saint Petersburg, and in 1869, he was invited to return to take up the post of Assistant Professor at the Mining Institute. This gave him ample opportunity to engage in research and, in particular, the first geological mapping of the Urals and European Russia. The magnitude of this enterprise is hard to imagine. The region he mapped was both vast and often remote, with complex geology to unravel. The building of new railway lines through previously isolated Russian territories provided him with access to these regions, as well as new outcrops created during the railway construction.

Alexander Karpinsky in 1897.
Photographer unknown

The mapping enabled the abundant mineral resources of this part of Russia to be delineated, and in turn, their exploitation helped power the Russian economy of the late 19th century. As a consequence, Karpinsky was made Imperial Director of Mining Research in 1885, and elected to the Russian Academy of Sciences in 1886. Following the Russian Revolution in 1917, he became the Academy's first elected president and held this post for the remainder of his lifetime.

Karpinsky's research was multifaceted, and he studied mineralogy, petrology, tectonics, paleontology and stratigraphy with equal enthusiasm. He was the first to describe the main features of the tectonic structure of the Russian Platform. Moreover, he integrated tectonics with an understanding of the areal distribution of sedimentary rock types to pioneer the concepts of paleogeography. He related the distribution of land and sea in past geological periods to tectonic uplift and subsidence events.

Karpinsky was also a "big picture" geologist. An 1888 paper, *Geology, geognosy, and palaeontology: on the regularity in outline, distribution, and structure of continents*, foreshadowed some of the concepts that would be pioneered by Alfred Wegener (although Karpinsky did not go so far as to suggest continental drift was possible) and matched Eduard Suess

Russian Imperial Academy of Sciences building, Saint Petersburg.

Helicoprion – Fossil specimen and modern reconstruction

Mining in the Ural Mountains in 1910.

in *Antlitz der Erde*, in terms of taking a pioneering global view of tectonics. Karpinsky often commented that *"the geologist needs the whole Earth"* — he was at the forefront of promoting the notion that any interpretation of the geological structure of the Earth requires a global outlook.

In 1898, Karpinsky became involved in an unusual paleontological diversion, a puzzle that still intrigues geologists today. A school inspector exploring the Ural Mountains encountered a strange fossil in some Permian strata. Whorl-shaped like an ammonite, it was clearly an arrangement of shark's teeth. A photograph and description were sent to Karpinsky for his comments. He named it *Helicoprion* (from the Greek *helico*, "spiral," and *prion*, "saw") and suggested that it initially protruded from the snout of the shark. Admittedly, this seems unlikely, so in subsequent interpretations, he decided the apparatus were not teeth, but rather defensive spines either on the tail or dorsal region of the shark. *Helicoprion* fossils have been found in some other locations outside of Russia and have puzzled paleontologists for over 100 years. It appears they are indeed teeth, a kind of circular saw in the mouth. However, because of how the teeth grow, only a portion of the whorl, with just a dozen or so teeth, would have been exposed at any time.

Karpinsky passed away in 1936. His ashes were placed within the Kremlin Wall in Moscow, a mark of the great esteem with which he was held for his contributions, not only to geological science, but to the location of mineral resources that helped develop the Russian economy. A number of geographic features carry his name, as does the Russian Geological Institute (VESGEI). VESGEI is the successor and custodian of traditions of the first state geological institution in Russia, the Geological Committee. Effectively the state geological survey, it was established in Saint Petersburg in 1882, by a decree of Emperor Alexander III (with guidance from Karpinsky) to study the geology of Russia. This included the compilation of a geological map of all Russian territories, as well as investigating the prospectivity of the country's regions for different mineral resources. Karpinsky was the first to chair this committee and the fact that its successor institute now carries his name is a fitting tribute to the great pioneer of Russian geology.

REFERENCES

This essay has drawn upon information from the following sources:

https://www.prlib.ru/en/history/618925

https://www.strangescience.net/karpinsky.htm

http://www.vsegei.ru/en

Ewing, S. 2017. *Resurrecting the Shark*. Pegasus Books Ltd.

Landscape in the North-West Highlands of Scotland

Ben Peach & John Horne

Sometimes, two scientists work so closely together and complement each other so well that their names become inseparable. So it is with Peach and Horne, names that are legendary in the context of the mapping of the North-West Highlands of Scotland and the gathering of evidence for thrust tectonics. Their 1907 memoir and associated mapping of the region set a standard that is still considered exemplary today. Generations of geologists have travelled to their study area to follow in their footsteps and train in the mapping of structural complexity, in the hope that they can emulate Peach and Horne's work, which, in the words of the great Austrian geologist, Edward Suess, *"rendered the mountains transparent."*

The North-West Highlands remains one of the few wilderness areas in Britain, famous for spectacular scenery where isolated mountains rise out of peat bog. To geologists, the region is equally famous for its spectacular geology, involving some of the oldest rocks on the planet in a stretch of outcrops 200 km long and 25 km wide.

In the second half of the 19th century a controversy had developed concerning the stratigraphy and structure of the region. The basic rock units present had been known since the early 1800s. In apparent stratigraphic order (lowest first) there are: a deformed set of gneiss and other metamorphic rocks (now termed the Lewisian Complex); a series of reddish sandstones (Torridonian Sandstone); white quartzites and limestones (now known to be Cambrian and Ordovician); and a thick series of metamorphosed sediments (the Moine Schists).

Ben Peach (right) and John Horne (left) outside the Inchnadamph Hotel whilst leading a field excursion in 1912.

John Macculloch, who first described the geology of the region, considered this a standard younging-upwards stratigraphic succession. This idea was adopted by the titan of mid-19th century geology, Sir Roderick Murchison, the 'King of Siluria,' with the Moine Schists being regarded as Silurian. The uppermost (Durness) limestones, beneath the Moine Schists, contained fossils that indicated a 'lower Silurian' age (Ordovician in the current sense) (the first fossils had been found by Ben Peach's father). Above the Moine Schists lay Devonian Old Red Sandstones. Murchison, therefore, reasoned that despite their metamorphosed state and the unmetamorphosed state of the sediments below them, the Moine Schists must be part of his beloved Silurian system.

The idea that the geology of the North-West Highlands was so simple sat ill with a number of geologists working in the region. The first was James Nicol, Professor of Geology at Aberdeen University, who had accompanied Murchison on one of his visits to the region. He gathered evidence showing multiple repetitions of the strata, and that the contact between the Moine Schists and the underlying rocks was tectonic in origin. Unfortunately, the prestige of Murchison, supported by his protégé, Archibald Geikie, the first Director of the Scottish Geological Survey, meant that Nicol's views were largely ignored.

Nonetheless, the doubts continued to amass. In 1883, Charles Lapworth, Professor at Birmingham University, who had recognised imbrication in the southern Uplands of Scotland by means of biostratigraphic control using graptolites, was advocating that the Moine Schists were tectonically emplaced over the underlying sediment. He noted the mylonites at the fault boundary, evidence for deformation by lateral translation of the overlying thrust sheet. By this time, Murchison had passed away and Geikie could hardly ignore the evidence that was gathering. Accordingly, he assembled a team led by two of his best geologists — Peach and Horne — to map the area and provide the definitive evidence.

Benjamin Peach was born in Gorran Haven, Cornwall in 1842. His father, Charles, was an amateur naturalist and geologist.

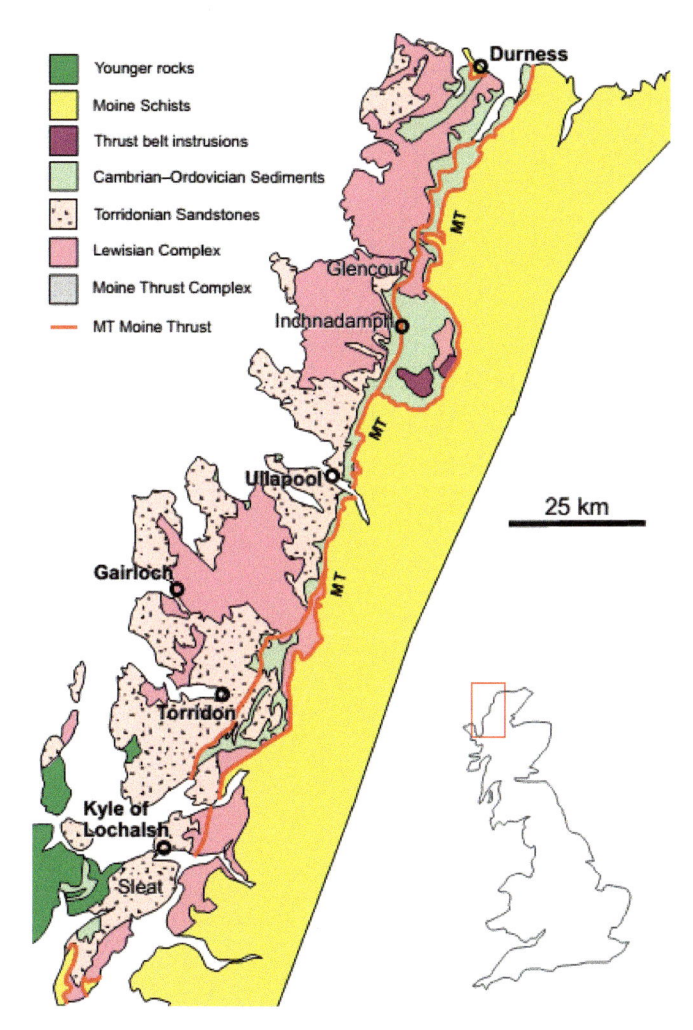

Great Geologists | 79

Geological map of the North-West Highlands region mapped by Peach and Horne. Adapted from Butler (2010).

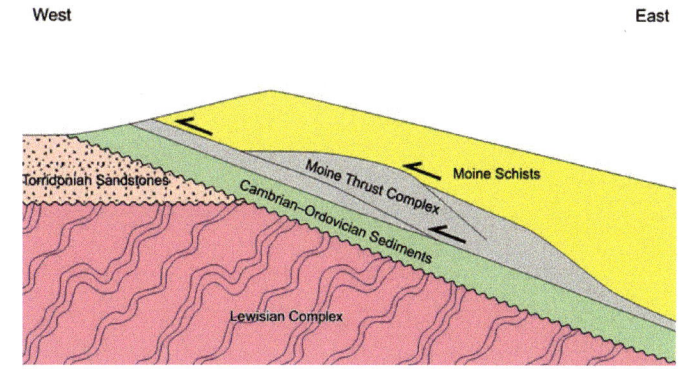

Simplified geological cross-section across the North-West Highlands showing the main rock units present and their relationship as understood today.

When posted to northern Scotland, working for the coastguard, he collected fossils from the Durness Limestone that aroused Murchison's interest in the geology of the region and, in effect, initiated the Highlands geological controversy. There is some irony that it was Ben Peach who contributed to its resolution — on Murchison's recommendation, Ben had been educated at the Royal School of Mines in London, and then joined the Geological Survey in 1862 as a geologist, moving to the Scottish branch in 1867.

John Horne was born on 1 January 1848, in Campsie, Stirlingshire. He was educated at Glasow High School and the University of Glasgow. He joined the Scottish Branch of the geological survey in 1867 as an assistant and became an apprentice to Ben Peach. The two soon became good friends and collaborators. Horne was a logical thinker and writer, complementing Peach's skills of creative thinking to resolve complex geological structures.

Peach and Horne worked on many various aspects of Scottish geology, including mapping in the Southern Uplands and in the coal belt of the Central Lowlands. Here, Peach took a special interest in the Carboniferous crustacean fossils he collected, later describing them in a major monograph. They also took geological holidays together that resulted in publications — on Orkney and Shetland, for example. But it was in the North-West Highlands that they made their greatest mark, unravelling the complex geological structures and producing a definitive map of the region.

Although the idea of horizontal movement along low-angle detachment surfaces had been discussed by Alpine geologists as early as 1841, in Britain, faults were largely considered as vertical features. It did not take Peach and Horne and their team long to realise that thrusting was the dominant structural style in the North-West Highlands (indeed, so rapidly was Geikie converted to this view, that it was he who introduced the term 'thrust plane'). By 1884, they had published their first paper to explain this, although a full description in the memoir accompanying the geological map took an additional 20 years to complete! Peach and Horne led a team of geologists that progressively mapped the region at the extremely detailed scale of 1:10,560 (six inches to the mile), beginning on the north coast in 1883, and completing the mapping on Skye in 1897. It took ten more years to write-up their seminal memoir, "*The Geological Structure of the North-West Highlands of Scotland.*"

The 1907 memoir is more than an exceptionally detailed description of regional geology. It explains how mountains form by the long-term deformation of the continental crust. It describes deformation structures at a variety of scales and explains how the recognition of these in the field can be used to understand broader tectonic processes. It is more than the first major synthesis of thrust belt structure — Peach and Horne provided the basis for understanding fault and shear zone processes, and methods for unraveling tectonic histories in metamorphic basement. It established the North-West Highlands as a key location for training in structural geology, a situation that remains unchanged today. Moreover, it provided an impetus for the understanding of tectonic processes in Europe, North America and beyond.

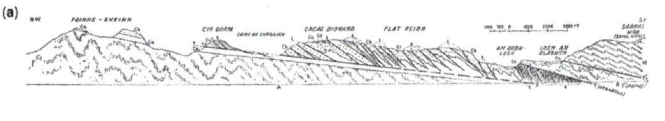

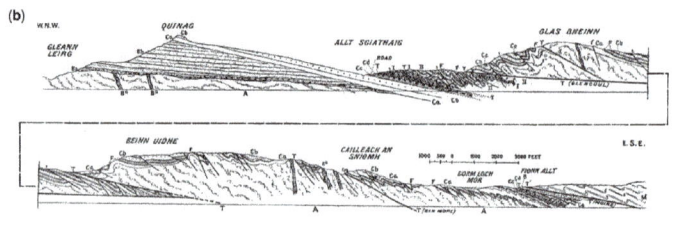

Structural cross-sections from the 1907 North-West Highlands Memoir.

The Arnaboll Thrust (part of the Moine Thrust Complex) at Ben Arnaboll with Lewisian Gneiss thrust over Cambrian quartzite. It was outcrops such as this that enabled Lapworth and Peach and Horne to appreciate the true nature of the rock succession in the North-West Highlands. Geikie introduced the term "thrust plane" after visiting this outcrop with Peach and Horne. Photograph courtesy of Prof. Rob. Butler, University of Aberdeen.

Peach and Horne were the leaders of a team, and each team member contributed to the research in their own way. For example, in 1889, one of the team members, Henry Cadell, developed a 'squeeze box' — a method of mixing layers of plaster of Paris with wet sand and clay in a box to which a horizontal force could be applied. This caused deformation of the 'strata' within it, mimicking some of the folding and thrusting seen in the North-West Highlands. This greatly helped Peach and Horne understand the processes that had led to the large westward translation of the Moine Thrust Sheet, by what we now know to be as much as 100 km.

The closing chapter of Peach and Horne's work on the North-West Highlands was leading a field trip for the British Association for the Advancement of Science in 1912. This was attended by 30 notable geologists from Britain and Europe, including the great Swiss structural geologist, Albert Heim. The party enjoyed an exceptional trip and passed the evenings composing and singing *La Chanson du Moine Thrust* (the Moine Thrust Song). They also paid extensive tribute to the brilliant work of their leaders. Heim remarked, "*They are a couple of scientists, Investigator-Twins, such as I never have seen before in my life, two men so delightfully developed in a wonderful common work of research.*"

The North-West Highlands is now a Geopark (www.nwhgeopark.com) with abundant opportunities for anyone with an interest in following in the footsteps of Peach and Horne. Geologists from across the world regularly visit to be amazed at the detailed mapping and unravelling of complex structural geology by one of the most famous pairings in the history of geological research.

Henry Cadell demonstrates his "squeeze box" for simulating the deformation of rock units in the North-West Highlands in a photograph from 1889.

REFERENCES

This essay has drawn upon information from the following sources:

Butler, R.W.H. 2010. The Geological Structure of the North-West Highlands of Scotland – revisited: Peach et al. 100 years on. In: Law, R.D., Butler, R.W.H, Holdsworth, R.E., Krabbendam, M. & Strachan, R.A. (eds.) *Continental Tectonics and Mountain Building: The Legacy of Peach and Horne.* Geological Society, London, Special Publications, 335, 7-27.

Dryburgh, P.M., Ross, S.M. & Thompson, C.L. 2014. *Assynt: The Geologists Mecca.* Edinburgh Geological Society, 36pp.

Goodenough, K. M. & Krabbendam, M. 2011. *A Geological Excursion Guide to the North-West Highlands of Scotland.* Edinburgh Geological Society, 215pp.

Mendum, J. & Burgess, A. 2017. John Horne (1848-1928). *The Edinburgh Geologist*, 61, 7-12.

Oldroyd, D.R. & Hamilton, B.M. 2002. Themes in the early history of Scottish geology. In: Trewin, N.H. (ed.) *The Geology of Scotland.* The Geological Society, London, 27-44.

Oldroyd, D.R. 1990. *The Highlands Controversy.* The University of Chicago Press. 438pp.

Rider, M. 2005. *Hutton's Arse.* Rider-French Consulting Ltd. 214pp.

http://edinburghgeolsoc.org/eg_pdfs/issue57_peach.pdf

Geodynamic reconstruction of the Earth during the mid-Cretaceous.

Alfred Wegener

Geodynamic reconstructions of the disposition of continents and oceans through geological time are now standard tools used by geologists. Using such industrial geologists can make predictive models of facies and mineral distribution, including source rocks and reservoirs, and we can better constrain tectonic events influencing the stratigraphic develooment of a region. Increasingly they can be used in modelling the climate of the past and to understand sediment flux from hinterland to depositional basin. Outside of practical considerations a sense of genuine wonder exists, even for the layman, to look at an overview of the Earth as it was millions of years ago – as close to a journey in a time machine as is possible. The genesis of this "time machine" that we now take for granted lies in the work of Alfred Wegener who, in the early part of the 20th century, was the leading proponent of the idea that in the geological past the continents were in a markedly different disposition to that seen today – the idea of continental displacement, more commonly known as *continental drift* and the forerunner of the modern science of plate tectonics. Wegener's story is all the more remarkable for the vehement opposition his theory generated in most of the geological community of the time and the fact that he was not a geologist by training, but primarily a meteorologist.

Wegener was born in 1880 in Berlin and proved to be an adept scholar and a man of action. Although his PhD was in astronomy, his main research interests were in atmospheric physics and climatology. He gathered data for his research from balloon ascents (setting a record for a continuous balloon flight of 52.5 hours in 1906) and from 1906 onwards made a number of expeditions to Greenland to gather data on polar climates and meteorology. The early part of his career was spent at the University of Marburg where by 1910 he had written and had published what became a standard textbook on the thermodynamics of the atmosphere.

It was whilst at Marburg that his scientific curiosity led him to initiate his investigation into continental drift. In early 1911 whilst examining a new atlas given to a colleague as a Christmas present, he was struck by the potential match of coasts of Africa and South America, especially at the edge of their respective continental shelves. Reading a paper by Erich Krenkel in July 1911 that mentioned the similarities of the Cretaceous geology between Brazil and Africa further piqued his interest, as did a paper on the "Former ice ages of Earth" by Konrad Keilhack that outlined the similarities in Carboniferous geology of the Southern Hemisphere continents. In rapid time he assimilated a good deal of geological data such that by January 6th, 1912, he was able to give a lecture on his developing theory to the German Geological Association in Frankfurt, followed by an article in the journal *Petermanns Geographischen Mitteilungen*. Despite serving in the German Army in the First World War, (and being twice wounded) by 1915 he had prepared a first version of his book summarizing his theory on *The Origins of the Continents and Oceans*. Four editions would appear, the last in 1929, and the 1922 edition, translated into several languages, would open up an often heated and vigorous debate within the geological community on the validity of the theories expressed within it.

Wegener was not the first to notice the apparent fit of the continents if moved from their present day position, nor was he the first to suggest that the continents had indeed moved. Abraham Ortelius, compiler of the first global atlas, noted in 1596 that the Americas had been *"torn away from Europe and Africa....by earthquakes and floods"*. In 1910 the American geologist Frank Taylor proposed that tidal forces generated by the capture of a comet that is now the moon caused continents to slide from the poles towards the equator. Unlikely as this theory is, he did correctly suggest that the Himalayas originate from collision between India and Asia and noted the importance of the mid-Atlantic ridge as suggestive of progressive separation of Africa and South America.

Wegener's critical contribution was to bring together all the evidence that suggested that the continents could have been conjoined around 300 million years ago into an original *"Urkontinent"*, or *Pangea*. Had the continents once been conjoined then their geology should match at the edges *"it is just as if we were to refit the torn pieces of a newspaper by matching their edges and check whether the lines of print run smoothly across"* he wrote. *"If they do, there is nothing left but to conclude that the pieces were in fact joined in this way"*. For Wegener, one of the most obvious examples of this was the distribution of late Paleozoic glacial deposits. Two examples of paleontological matching across continents were cited. First were fossils of the small Permian reptile, *Mesosaurus*, found in identical rocks in south-eastern Brazil and south-western Africa. Second was the fossil fern *Glossopteris*, found across the Southern Hemisphere. To Wegener their distribution suggested continental juxtaposition.

Professor Alfred Wegener Circa 1924. Photographer unknown.

As well as gathering empirical evidence to support his theory of moving continents he also understood that isostasy implies a different nature of oceanic and continental crust and that this crust must lie on a material that has fluid properties. Therefore *"if...the continental blocks really do float on a fluid... there is clearly no reason why their movements should only occur vertically and not also horizontally"*. He supposed that the driving mechanisms might be the centrifugal force of the Earth's rotation (*"Polflucht"*) or astronomical precession, but with some degree of foresight noted that *"it is probable....that the complete solution of the driving forces will still be a long time coming"*.

To say that the geoscience community of the 1920's was skeptical about the idea of continental drift would be an understatement. Land bridges across oceans were invoked to explain paleofaunal and paleofloral similarities between continents, or faunal and geological similarities were simply denied to exist. Geophysical arguments were presented to deny the possibility of crustal movement.

Most telling was that the theory that Wegener suggested for continental movement was (rightly) deemed impossible. Consequently the observations that Wegener had summarized that were strongly suggestive of a supercontinent assembly were dismissed for the want of a theory to explain them.

Not all eminent geologists of the time were opposed to Wegener's ideas. The South African Alexander du Toit published a comprehensive favourable comparison of South American and African geology in 1927 and published *Our Wandering Continents* in 1937. The British geologist Arthur Holmes provided an illustration of how convention currents in the mantle could drive continental movement in 1929, an argument that Wegener himself had advanced in later editions of his "*Origins...*" book. To quote Holmes *"granted convection currents the continents may open out and reclose in an endless pattern of varieties."*

Nonetheless many of the leading geologists of the time, especially Americans such as Charles Schuchert, Edward Berry (an article entitled "*Germanic Pseudo-Science*" gives a flavour of his sentiments) and Bailey Willis, were "*united in a crusade against mobilism*" and remained staunchly against continental drift for the entirety of their careers. It would take the 1960's development of plate tectonic theory drawn from oceanographic and paleomagnetic observations to convince geologists that Wegener had been right in his hypothesis that the continents were not fixed.

Wegener died not long after his 50th birthday in 1930 when participating in his fourth Greenland expedition. After delivering supplies by dog-sled to the Eismitte research station near the centre of the Greenland ice sheet, he suffered an apparent heart attack on the return journey on skis to the west coast. His body was buried by his Inuit companion (who himself perished soon after) and discovered the following year. At the request of his widow, his grave remains on Greenland, the location of some of his most important meteorological research.

What is striking about Wegener's geological work is the holistic approach he took. For him, *"the forces which displace continents are the same as those which produce great fold-mountain ranges. Continental drift, faults and compressions, earthquakes, volcanicity, transgression cycles and polar wandering are undoubtedly connected causally on a grand scale."* No geologist today would argue with the sentiment of this conclusion. We owe a great debt to Alfred Wegener for directing geological thought to this bigger picture.

REFERENCES

This essay has drawn upon information from the following sources:

Frankel, H.R. 2012. *The Continental Drift Controversy. Volume I: Wegener and the Early Debate.* Cambridge University Press. 604pp.

Gohau, G. 1990. *A History of Geology.* Rutgers University Press. 259pp.

Greene, M.T. 1982. *Geology in the Nineteenth Century.* Cornell University Press. 324pp.

Greene, M.T. 2015. *Alfred Wegener.* John Hopkins University Press. 675pp.

Hallam, A. 1973. *A Revolution in the Earth Sciences.* Clarendon Press, Oxford. 127pp.

Hallam, A. 1983. *Great Geological Controversies.* Oxford University Press, 244pp.

Lawrence, D.M. 2002. *Upheaval from the Abyss.* Rutgers University Press, 284pp.

Le Grand, H.E. 1988. *Drifting Continents and Shifting Theories.* Cambridge University Press. 313pp.

Molnar, P. 2015. *Plate Tectonics: A Very Short Introduction.* Oxford University Press, 136pp.

Oldroyd, D.R. 1996. *Thinking About the Earth.* The Athlone Press, 410pp.

Oreskes, N. 1999. *The Rejection of Continental Drift.* Oxford University Press. 420pp.

Powell, J.L. 2015. *Four Revolutions in the Earth Sciences.* Columbia University Press, 367pp.

Arthur Holmes

One of the distinguishing traits of geologists is the ability to discuss Earth's history in millions (indeed billions) of years in a very matter of fact way. Laypersons often find this amazing – "those limestones are 400 million years old?" they will ask in slightly disbelieving tones when during a car journey you have casually remarked on an outcrop you are driving past and commented on the age of the rocks present. Moreover, "how do you know?" will be their next question. A more complex question than they might imagine of course, but part of the answer lies in the pioneering work of English geologist Arthur Holmes with whom the numerical dating (geochronology) of the standard geological time scale as we know it today began. But Holmes achieved much more in his career than providing chronological ages to the subdivisions of geological time. He suggested a mechanism for explaining plate tectonics at a time when this process was viewed as heretical by most of the geological establishment of the mid-20th century. He also authored what was the standard university geology textbook of much of the late 20th century: *"Principles of Physical Geology"*. Every British geology undergraduate of the 1950s through to the 1980s can recall swatting up on the knowledge and theory presented in their copy of *"Holmes"* in preparation for their exams. It was widely used internationally and translated into several different languages.

The Scottish Enlightenment geologist James Hutton had, by the latter part of the 18th century, recognised 'the abyss of time'. Geological processes and the rock record implied a much, much longer duration of Earth history than estimates based purely on interpretations of religious texts. Building on the work of William Smith and Georges Cuvier, stratigraphers of the 19th century (Murchison, Sedgwick, d'Orbigny and others) had set about creating the subdivisions of geological time as expressed by intervals of the rock record characterised by particular fossils placed in order by the Law of Superposition (Silurian, Devonian, etc). But what was the numerical age and duration of these subdivisions and of Earth history in general? As the 20th century dawned these questions could not be answered with certainty. The situation was akin to knowing history but without any dates.

Lord Kelvin, the Scottish mathematician, and physicist, William Thompson had provided a typically quoted view of the age of the Earth in the late 18th century based on thermodynamics. His view that the Earth was 20 million years old (although his earlier estimates allowed for it to be as much as 400 million years old) was based on the time it would take for the Earth to cool from its assumed molten state at formation to its present day state. For many geologists working as Kelvin's contemporaries, the notion of a 20 million year old Earth seemed too short to account for all the physical processes that the rock record indicated had occurred in the geological past. Using estimates of the time to accumulate sedimentary thickness, many geologists preferred to estimate the age of the Earth as around 100 million years, although this was a matter of some dispute.

Arthur Holmes entered into the debate on the age of the Earth in the pre-war years of the 20th century. These were years when discoveries concerning radioactivity by Henri Becquerel, Marie and Pierre Curie, Ernest Rutherford and others were

Arthur Holmes aged 22, the time at which he was first publishing on the numerical age of the Earth.

changing the understanding of physics and ultimately geology at a rapid pace. The energy provided by the radioactive decay of uranium and other elements indicated that the earth was not simply cooling as Lord Kelvin had suggested. Moreover, the time for the transformation of radioactive elements from unstable parent elements to stable daughter elements meant that the proportion of one to another could be used to measure numerical age (radiometric dating). As early as 1906 Ernest Rutherford had suggested that a rock could be 500 million years old based on such an analysis.

Holmes was born in 1890 in Gateshead in north-eastern England. Inspired by teaching at school, he gained a scholarship to study physics at the Royal College of Science (now part of Imperial College) in London. He became increasingly interested in geology whilst at university and the link between physics and geology that research into radioactive elements represented - in particular the radioactivity studies carried out in the physics department by Professor Robert Strutt. Holmes took on a research project with Strutt to use radiometric dating to date a set of rocks from different geological periods, the oldest being a speciment from the Precambrian from Sri Lanka. Rutherford and others were using the measurement of helium, which was understood to be released in the transformation from parent to daughter element, as the basis for their radiometric dating. Because some helium is lost during the radioactive process, its results were likely to only provide minimum ages. Therefore, inspired by the pioneering studies of the American chemist Bertram Boltwood, Holmes decided to measure the quantity of uranium versus the amount of lead (the ultimate stable daughter element derived from a parent unstable element). This was painstaking, difficult work but ultimately he was able to derive a set of ages. This included the observation that the oldest Precambrian sample was 1640 million years old. These results were presented to the Royal Society (by Strutt on behalf of his student) and subsequently published in the society's Proceedings. It seems very commendable that Strutt took no direct credit or co-authorship for the work of his student!

The reason why Holmes did not present the results of his work was that at the time he was engaged in an expedition to Mozambique on behalf of a mineral mining company. His principal motivation for participation was financial, his university scholarship being barely enough to keep pace with the cost of living, but nonetheless, he enjoyed the opportunity to carry some genuine geological exploration. It almost cost him his life. Shortly before the expedition ended he contracted

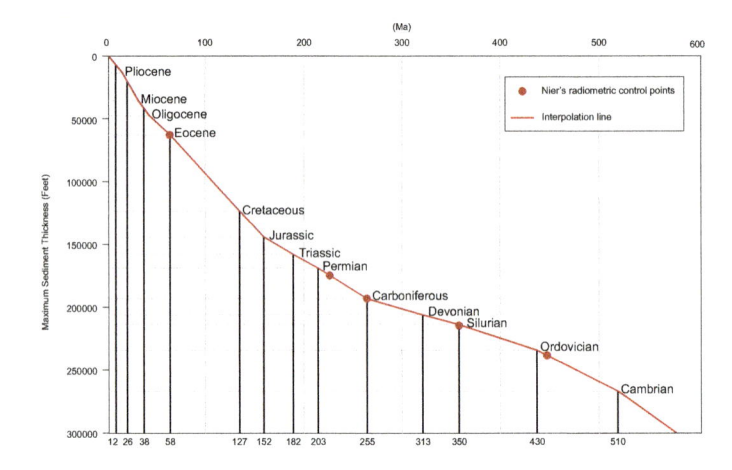

Holmes' 1947 interpolation of numerical ages (derived from the work of Alfred Nier) with sedimentary thickness to generate a Phanerozoic timescale.

the complication of malaria known as blackwater fever. The nuns in charge of the local hospital where he was admitted did not expect him to live and a premature notice of his death was sent back to London.

But survive he did and he returned to London to continue his studies into the radiometric dating of rocks. By 1913 he had published a short book on the subject, *"The Age of the Earth"*, subsequently revised several times in his lifetime. His motivation for doing so was the increasing and perhaps surprising resistance in the geological community to the age of the Earth being suggested by Holmes and others using radiometric techniques. Whereas they had previously considered Lord Kelvin's calculation of the age of the Earth of 20 million years to be too short, they now considered radiometric ages of 1600 million years to be too long! Put simply, many geologists felt that there was not enough rock on the planet to satisfy such a long Earth history. Plate tectonics and the true nature of our dynamic Earth would ultimately provide the explanation of this conundrum, but in the meantime Holmes felt the need to explain that geologists need not doubt the physics and chemistry that radiometric dating is based upon.

His physical condition after the malarial infection contracted in Mozambique excused him from front line service in the Great War. Instead he was engaged in mapping and mineral studies to aid the war effort whilst trying to continue his research. In particular, he needed to adapt the developing knowledge of

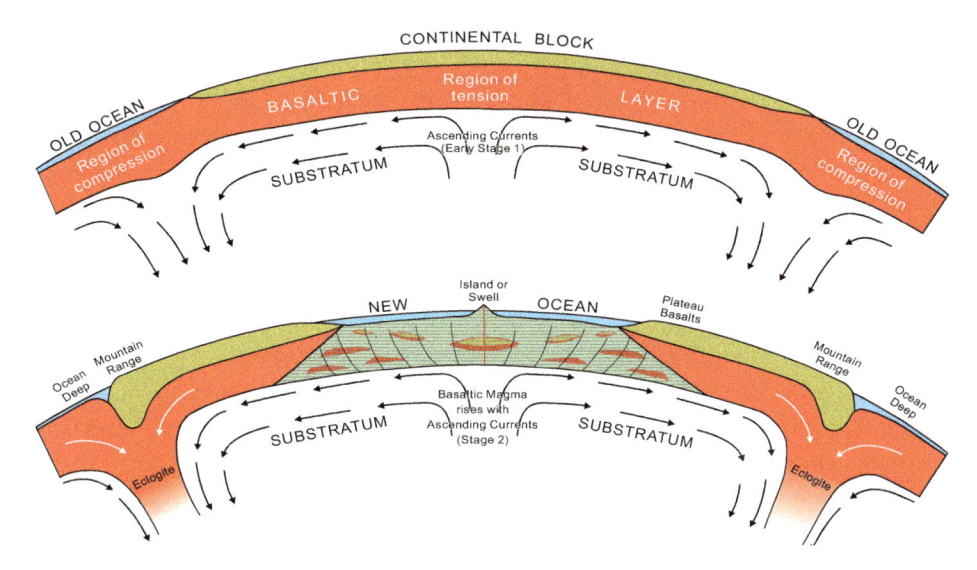

A representation of the mechanism by which Holmes envisaged continental drift might operate, first presented in Holmes' *Principles of Physical Geology* in 1944.

isotopes as pioneered by the work of Frederick Soddy who had shown that not all lead present in rocks was derived from radioactive decay. Uranium decayed to ^{206}Pb, thorium decayed to ^{208}Pb, and ^{207}Pb was at the time believed to be of non-radiogenic origin. It thus became necessary to determine the proportions of each isotope, which in a time before mass spectrometers, required measurement of their atomic weights. Unfortunately, following some initial research, the war prevented access to the laboratories at the Radium Institute in Vienna where such work was carried out.

Following the war, Holmes, despite his growing reputation in geological circles and investigation of radiometric dating using isotopes of lead, was unable to get a permanent post at a British university. This led him in 1920 joining the oil industry as the Chief Geologist of the Yomah Oil Company, active in Burma (modern-day Myanmar). This was both an unsuccessful and tragic adventure – his three year old son died after contracting dysentery. No major oil field was located in the company's acreage and the continuously precarious finances of the company led to a generally unsatisfactory life. Not surprisingly, Holmes returned to Britain in 1922 and for two years took on various jobs including running a shop selling 'oriental crafts' in Newcastle.

In 1924 his luck changed and he was appointed Head of the newly-formed Geology Department at Durham University (initially he was the only staff member) and this allowed him to resume his geological research. In particular he was engaged in the committee set up by the National Research Council in America for the "Measurement of Geologic Time by Atomic Disintegration". This required reviewing the growing numbers of radiometric analysis now being published. Few met his high standards, with the results ultimately published in 1931 with the lead author being the American mineralogist Albert Knopf. His search for more reliable radiometric methods led him to contemplate how the original isotopic ratios of igneous rocks could be determined, which in itself opened up a new avenue of research of igneous petrogenesis. Collaboration with William Urry of the Massachusetts Institute for Technology suggested that a return to helium measurements might prove fruitful, but the results were shown to be spurious. It was not until Alfred Nier demonstrated the true nature of the isotopes of uranium and lead that reliable results began to emerge and these results were incorporated in the 1947 version of Holmes' geological timescale which interpolated ages for the geological periods between the radiometric control points using sediment thickness as a guide to duration. Holmes himself knew that there were flaws in this method, but it did provide the geological community with numerical age values for the whole of the Phanerozoic.

During World War II Holmes was charged with the rapid education of RAF cadets. Realising that no textbook existed that captured the latest geological thinking he set about creating one from his lecture notes. The result was the classic *Principles of Physical Geology* first published in 1944. This beautifully written and illustrated book became the standard university textbook in its first and subsequent editions. Widely read, it inspired future generations of geologists.

One aspect of the first edition is particularly remarkable – it makes reference to continental drift, the argument promoted by Alfred Wegener that the continents had not always been in their current position and had moved through geological time. Although now fully explained by plate tectonic theory, the concept of moving continents was viewed by many geologists in the mid-20th Century with extreme scepticism. Holmes had been contemplating this problem since the late 1920's and considered that differential heating of the Earth's interior, generated by the decay of uranium and other radioactive elements, caused convection in the "substratum" (the mantle in today's terminology), on which the continents floated. Convectional cells would be generated that could drag continents sidewards and apart allowing new crust to rise up and form. First presented to the Edinburgh Geological Society in 1927, this theory is a brilliant forerunner of the mechanisms by which it is now known that plate tectonics work. Some contemporary geologists were less impressed, as this quote from William Bowie shows: *"Holmes brings out a new thought which is even more impossible than Wegener's. That is that the submerged ridge through the Atlantic Ocean is the place at which North and South America separated from Europe and Africa… I believe that we need to apply elementary physics and mechanics to the continental drift problem in order to show how impossible drifting would be"*. Holmes always considered his work in this sphere highly theoretical and was never absolutely convinced of its correctness. Nonetheless it is to his great credit that he included it in his teaching materials, to be ultimately vindicated by the geophysical observations of the 1960's and onwards.

Holmes would continue to synthesise radiometric data and evolve the geological timescale until his death in 1965. This incorporated the age of the Earth determined as 4550 million years +/- 70 million years as determined by Claire Patterson in 1953 integrating data from meteorites determined as forming at the same time as the Earth. This age has not fundamentally changed for over 60 years. Arthur Holmes was always the impetus for such developments and the critical analysis of their validity. In this respect he can be considered as the "father" of geological timescales, that is to say the integration of geological ages with absolute ages - geochronology.

The contributions of Holmes to geology are thus remarkable – the dogged pursuit of the numerical dating of the geological record, from which the rates of geological processes can be measured. This interest in process is linked to his promotion, against the conventional wisdom of the time, of the causes by which continental drift (and subsequently plate tectonics) might be explained. And ultimately he was able to convey the excitement of the changing face of geological understanding to generations of students through his masterly textbooks and papers. For many he is the greatest British geologist of the 20th century.

REFERENCES

This essay has drawn upon information from the following sources:

Hallam, A. 1973. *A Revolution in the Earth Sciences*. Clarendon Press, Oxford. 127pp.

Hallam, A. 1983. *Great Geological Controversies*. Oxford University Press, 244pp.

Le Grand, H.E. 1988. *Drifting Continents and Shifting Theories*. Cambridge University Press. 313pp.

Lewis, C.L.E. 2000. *The Dating Game*. Cambridge University Press.

Lewis, C.L.E. 2001. Arthus Holmes' vision of a geological time scale. In: Lewis, C.L.E. & Knell, S.J. (eds.) *The Age of the Earth: from 4004 BC to AD 2002*. Geological Society, London, Special Publications, 190, 121-138.

Lewis, C.L.E. 2002. Arthur Holmes unifying theory: from radioactivity to continental drift. In: Oldroyd, D.R. (ed.) *The Earth Inside and Out: Some Major Contributions to Geology in the Twentieth Century*. Geological Society, London, Special Publications, 192, 167-184.

Oreskes, N. 1999. *The Rejection of Continental Drift*. Oxford University Press. 420pp.

Powell, J.L. 2015. *Four Revolutions in the Earth Sciences*. Columbia University Press, 367pp.

Milutin Milanković

Milutin Milanković photographed whilst a student in Vienna.

That climate changes naturally through time and on different time-scales has long been understood. 19th century geologists were well aware that Earth's past climates were different from today. Initially, global climates were viewed as changing on a long term basis (the Cretaceous period being warmer than today for instance), but with the recognition by Louis Agassiz and others of ice ages in the relatively recent past, it became apparent that climate change could also be short-term. How quickly climate might change and the associated driving mechanisms were uncertain, although it was suspected that variations in Earth's orbital parameters creating changes in solar insolation (the amount of energy received from the Sun) might be at least in part responsible.

This conundrum was effectively solved in the first half of the 20th century by the outstanding Serbian scientific polymath, Milutin Milanković (often spelt in anglicised form as Milankovitch). Laborious calculations permitted Milanković to determine that cyclic variations in Earth's orbital parameters could profoundly control climate on scales ranging from tens of thousands to hundreds of thousands of years. Milanković's work was challenged because age-dating of rocks was in its infancy at the time he was publishing his ideas, meaning they could not be tested. Furthermore, the changes in insolation that he proposed as the driving mechanism for climate change were considered too small to be significant.

However, in the 1970s his work was rediscovered by paleoclimatologists and validated by data from deep-sea sediment cores and cores from ice sheets. It now forms a central pillar of our understanding of Earth's past climates and is used to explain some of the cyclicity seen in the sedimentary record. In turn, this cyclicity can be used to develop high-resolution timescales. Without an understanding of 'Milankovitch Cyclicity' our knowledge of Earth system science would be significantly poorer.

Milanković was born in 1879 in Dalj, a village on the banks of the Danube, in what is now Croatia. An able student, he moved to Vienna in 1896 to study civil engineering at the Vienna University of Technology. He graduated in 1902 and, after a year of military service, returned to the university to study for a PhD entitled *The Contribution to the Theory of Pressure Curves*. His PhD was quickly obtained before the end of 1904 and he then embarked on a career in construction engineering. Whilst working on practical dam, bridge and aqueduct design using reinforced concrete, he published a number of key theoretical papers alongside patenting some of his ideas. Such was his reputation that in 1909 that he was offered the Chair of Applied Mathematics at the University of Belgrade.

The move to academia was to signify a change in interests for Milanković with a focus on fundamental research. A topic that soon captured his attention was that of Earth's climate and the controls acting upon it. He noted that much of meteorology and climatology was "*nothing but a collection of innumerable empirical findings, mainly numerical data, with traces of physics used to explain some of them*". He decided to employ advanced mathematics, in a manner not previously attempted.

Milanković's '*Contribution to the mathematical theory of climate*', published in 1912, described the present climate on Earth and how the Sun's radiation determines the temperature on Earth's surface after passing through the atmosphere. He built upon this with '*Distribution of the Sun's radiation on the Earth's surface*' published in 1913. He calculated the intensity of insolation and developed a mathematical theory describing Earth's climate zones.

His ultimate aim was a mathematically-based theory which connected the thermal regime of Earth to variations in its orbital parameters. The theory could be applied to the geological past, especially the onset of ice ages. He wrote: "*such a theory would enable us to go beyond the range of direct observations, not only in space, but also in time… It would allow reconstruction of the Earth's climate, and also its predictions, as well as give us the first reliable data about the climate conditions on other planets.*" A paper published in 1914 ('*About the issue of the astronomical theory of ice ages*'), was a preliminary statement on the topic, noting that others, including the French mathematician, Joseph Adhémar, and the Scottish scientist, James Croll, had attempted to address the problem in the previous century.

Unfortunately for Milanković, his work was disturbed by political events. On 14th June 1914, Milanković married Kristina Topuzovich and went on honeymoon for the summer to his native village of Dalj, which was then in Austro-Hungary. Whist the newly-wedded Milanković was staying in Dalj, Archduke Franz Ferdinand, the heir to the Austro-Hungarian throne, was murdered in Sarajevo by a Serbian extremist. The consequence was a declaration of war by Austro-Hungary on Serbia, which in turn led to World War I. Milanković was arrested as a Serb and interned. Fortunately he was allowed to spend his captivity in Budapest and work in the library of the Hungarian Academy of Science, and the Central Meteorological Institute. His work

Milanković and his ideas celebrated on a Serbian stamp from 2004.

during this period mostly focused on calculating the climate on the inner planets of the solar system and of the Earth's moon, although he continued his research into the drivers of ice ages.

In 1919 he returned to the University of Belgrade and focused on the climate of the Earth both past and present. His main task was to produce a mathematical theory of insolation, which was summarised in a book published in 1920 entitled '*Mathematical Theory of Heat Phenomena Produced by Solar Radiation*'.

Encouraged by the Russian climatologist Wladimir Köppen and his son-in-law Alfred Wegener (of continental drift fame), he began calculations of variations in Earth's insolation over the last 650,000 years. Milanković spent 100 days without break completing the calculations using only pen and paper and prepared a graph of solar radiation changes at geographical latitudes at 65° north. Milanković believed that this latitude was most sensitive to a change of thermal balance. At 65° north, ice sheets develop not because it gets cold in winter but because it remains cool in summer. In this way major ice sheets can develop through a positive feedback, whereby the albedo of the snow- and ice-covered surface reflects away radiation coming from the Sun.

We know that it gets colder at higher latitudes in winter each year because the amount of sunlight reaching the Earth's surface decreases, due to the tilt of the Earth's axis. Milanković expanded this concept to larger timescales and three key cyclic variations in Earth's orbit around the Sun creating variations in insolation.

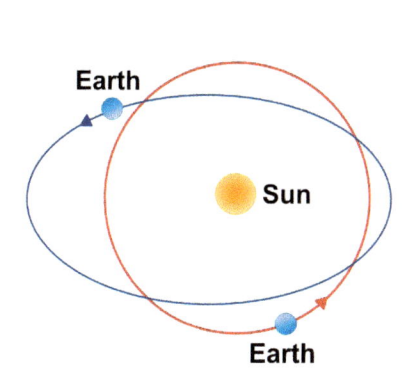

Eccentricity – Earth's orbit around the Sun varies from circular to elliptical on a cyclicity of c. 100,000 years

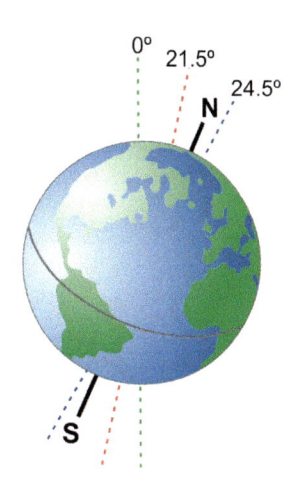

Obliquity – The angle of Earth's axis of rotation cycles between 21.5° and 24.5° with a periodicity of c. 41,000 years

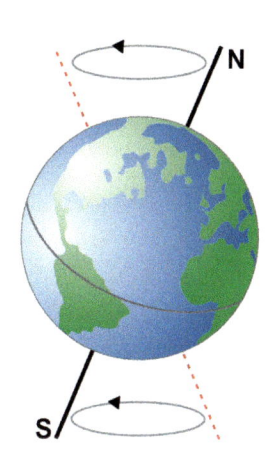

Precession – Earth's axis of rotation wobbles with a periodicity of c. 23,000 years

The key orbital parameters that make up Milankovitch cyclicity.

First, the shape of the Earth's orbit around the Sun changes periodically, being sometimes more circular and at other times more elliptical. The degree of ellipticity of Earth's orbit around the Sun is known as its eccentricity. A more elliptical orbit has high eccentricity. The length of one eccentricity cycle is about 100,000 years. Extreme variations in eccentricity occur with a cycle time of 405,000 years. At its most elliptical, the extra distance from the Sun can cut the amount of insolation by as much 30% compared to when the Earth and Sun are at their closest. For this reason, Milanković considered eccentricity of prime importance compared to the other two factors. Recent studies have noted that the 405,000 year eccentricity cycle is very stable and can be detected in the rock record for at least the last 250 million years. Driven by the gravitational interactions between Jupiter and Venus, it behaves as a metronome, forming one of the primary drivers of cyclicity seen in sedimentary sequences. Recently, 'grand cycles' with a periodicity of 1.2 million years and 2.4 million years have been recognised, the result of interactions between Earth and Mars.

Second, the angle or obliquity of the Earth's axis of rotation changes periodically. Today this angle is 23.5°, but it cycles between 21.5° and 24.5° with a periodicity of about 41,000 years.

Third, Earth's axis of rotation wobbles like a spinning top, so that axis of rotation draws out a cone over time, giving rise to a pattern of variation known as precession. This has a periodicity of around 23,000 years. The main impact comes from the fact that the seasons occur at different points on the eccentric orbit, changing the lengths of summer and winter.

The harmonic interference between the different periodicities of cyclicity was understood by Milanković as being critical in driving insolation changes and thus climate variations through time.

Milanković argued that these cycles would control the waxing and waning of polar ice sheets and be a primary control on the Earth's climate fluctuations. These ideas were vindicated in the 1970's when oxygen isotope records (a proxy for paleotemperature) from deep-sea sediment cores demonstrated cyclicity in keeping with Milankovitch's periodicities. Subsequent to this, studies on ice cores from the Antarctic that show temperature and atmospheric greenhouse gases (as determined from gas bubbles within the ice cores) vary with a periodicity completely in accord with the orbital cyclicity noted by Milanković.

In 1939, to collect his scientific work on the theory of orbitally-driven variations in insolation and their impact on past climates

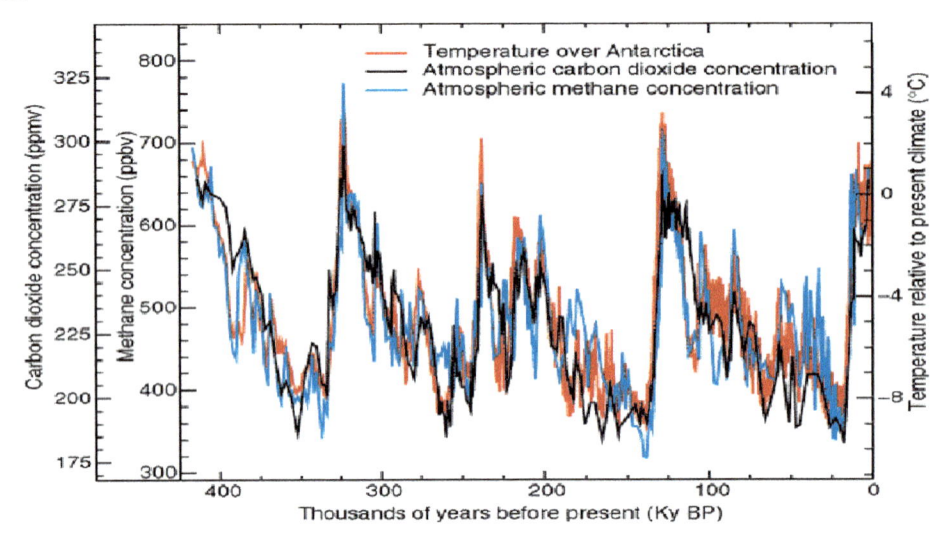

Temperature and greenhouse gas records from the Vostok ice core from Antarctica display variations in keeping with Milankovitch cyclicity and correspond to phases of polar ice sheet growth and decline.

into one volume Milanković began work on his *"Canon of Insolation of the Earth and Its Application to the Problem of the Ice Ages"*, which covered his nearly three decades of research, and summarized the universal laws through which it was possible to explain cyclical climate change and attendant ice ages.

Milanković spent two years arranging and writing the *"Canon"*. The manuscript was submitted to print on 2 April 1941 – four days before the attack of Nazi Germany and its allies on the Kingdom of Yugoslavia. In the bombing of Belgrade on 6 April 1941, the printing house where his work was being printed was destroyed; fortunately a printed copy of the manuscript had already been transferred to the printer's warehouse. After the successful occupation of Serbia on 15 May 1941, two German officers with backgrounds in geology visited Milanković. He gave them the only complete printed copy of the *"Canon"* for safekeeping in the University of Freiburg. The book was eventually published by the Royal Serbian Academy in German as *"Kanon der Erdbestrahlung und seine Anwendung auf das Eiszeitenproblem"*, a landmark in paleoclimate studies.

During the German occupation of Serbia from 1941 to 1944, Milanković withdrew from public life and decided to write a personal history going beyond scientific matters. This autobiography was published after the war, entitled *"Recollection, Experiences and Vision"* in Belgrade in 1952.

Milanković had a long-standing interest in popularising science. Between 1925 and 1928 Milanković wrote *Through Distant Worlds and Times* in the form of a series of letters to an anonymous woman. The work discusses the history of astronomy, climatology and science via a series of imaginary visits to various points in space and time by the author and his unnamed companion, encompassing the formation of the Earth, past civilizations, famous ancient and renaissance thinkers and their achievements, and the work of his contemporaries. In the "letters", Milanković expanded on some of his own theories on astronomy and climatology, and described the complicated problems of celestial mechanics in a simplified manner.
After the Second World War he began publishing numerous books on the history of science, including *Isaac Newton and Newton's Principia* (1946), *The Founders of the Natural Sciences: Pythagoras – Democritus – Aristotle – Archimedes* (1947) and *Techniques in the Ancient Times* (1955).

Milanković suffered a stroke and died in Belgrade in 1958. He is buried in his family cemetery in Dalj. His work has left a profound legacy for Earth sciences. Not only can it be used to explain some of the key driving mechanisms of climate change in the geological past, but Milankovitch cyclicity is increasingly recognised through its visible expression in the rock record. For example, the origin of rhythmically bedded successions, such as the Early Jurassic open marine limestones and marls of southern

Britain, have long been debated. The application of time-series analysis to a range of properties of the sedimentary succession (for example, percentage carbonate or organic carbon) has demonstrated that they can be explained in terms of the various Milankovitch cycles.

Moreover, if Milankovitch cyclicity can be recognised in the rock record, it can be used to "orbitally tune or "astronomically tune" the geological timescale. The geological timescale classically relies on radiometric dating. Unfortunately, reliable radiometric ages are not available for every stage or period. Therefore, recognition of orbitally-forced (Milankovitch) cyclicity can be used to estimate the duration of a period of geological time between two established datums. This technique is now a major contributor to updating the geological timescale.

Why Milankovitch cyclicity should be present in the rock record during periods considered to have minimal polar glaciation (parts of the Mesozoic, for example) is something of a mystery. One possibility is that glaciation is more frequent in the geological past than often considered. Alternatively, it may be that Milankovitch cyclicity affects the rock record by, for example, increasing and decreasing the amount and intensity of precipitation over a particular landmass. This in turn influences run-off which changes clay content of marine rocks and drives productivity changes and the deposition of organic matter, leading to a cyclic depositional pattern.

A further uncertainty is how far back in geological time the celestial mechanics that drive Milankovic cyclicity can be deemed valid. Milankovitch cyclicity has readily been recorded from Carboniferous cyclothems and has even been described from banded ironstone formations formed in the Precambrian.

What cannot be debated is that Milankovitch's work continues to resonate through geology, both providing answers and asking important questions.

REFERENCES

This essay has drawn upon information from the following sources:

Frakes, L.A., Francis, J.E. & Syktus, J.I. 1992. *Climate Modes of the Phanerozoic.* Cambridge University Press, 274pp.

Grotzinger, J.P. & Jordan, T.H. 2014. *Understanding Earth (Seventh Edition).* W.H. Freeman & Company, 672pp.

Hinnov, L.V. & Hilgen, F.J. 2012. Cyclostratigraphy and Astrochronology. In: Gradstein, F.M., Ogg J.G., Schmitz, M.D. & Ogg, G.M. (eds.). *The Geologic Time Scale 2012.* Cambridge University Press, 63-83.

Imbrie, J. & Imbrie, K.P. 1979. *Ice Ages: Solving the Mystery. Macmillan, 224pp.*

Nield, T. 2007. *Supercontinent: Ten Billion Years in the Life of our Planet.* Granta Books, 288pp.

Weedon, G.P. 2003. *Time-Series Analysis and Cyclostratigraphy.* Cambridge University Press, 259pp.

http://www.teslasociety.com/milankovic.htm

https://www.youtube.com/watch?v=vCzDAm1nYlI

Rhythmic bedding (light grey limestone, dark grey marls) in the Blue Lias (Early Jurassic) at Lyme Regis in southern England. Such bedding patterns can be interpreted as being the product of Milankovitch cyclicity.

Amadeus Grabau

Portrait of Amadeus Grabau from 1946.

China is now a powerhouse of geoscience research, but one-hundred years ago that was far from the case. In contrast to the intense research activity in Europe and America, geology was barely taught at Chinese universities and there was only an embryonic geological survey. Overall, the geological profession was held in low regard.

That changed with the arrival of an American geologist at Peking University in 1920, Amadeus William Grabau. Not only did Grabau revolutionise the status of geology in China and highlight the country's geology to the world, he also tied the observations he made there with his knowledge of geology in Europe and America to create a new global synthesis. Years ahead of his time, he recognised the importance of a dynamic earth with mobile continents and global sea-level change. Grabau can rightly be considered an important pioneer of sequence stratigraphy, eustasy, paleogeography and sedimentology.

Grabau was born in the town of Cedarburg, Wisconsin, in 1870, to a family with a long history of Lutheran ministry. When he was 15, the family moved to Buffalo, New York, where he became involved with the local Society for Natural History and the allied Agassiz Association (named after Louis Agassiz) which promoted field trips in order to 'study nature, not books'. Botany was his first interest, but Grabau soon began collecting fossils, especially trilobites and other Devonian fossils, from the now celebrated locality of 18 Mile Creek on the south side of Lake Erie. An evaluation of the stratigraphy and fauna of this area formed his Bachelors thesis at Massachusetts Institute for Technology (MIT) and was subsequently published as an extended field guide by the Buffalo Society for Natural History (1898, 1899). Still regarded as a classic of its kind today, the guide was republished in 1994.

Grabau extended his research to the stratigraphy of the Niagara Falls area and New York state in general and was accepted at Harvard University to undertake a PhD on the evolutionary history of the mollusc *Fucus*. By the early 1900s, he had established himself as a researcher who 'got things done,' first as a professor at Rensseler Polytechnic and then at Columbia University.

These were extremely productive and happy times for Grabau. He rapidly established himself as one of the leading Paleozoic specialists in North America and met his wife, Mary Antin, during the course of a field trip for members of the general public interested in natural history. Mary, a Russian émigrée, was to become a celebrated figure in

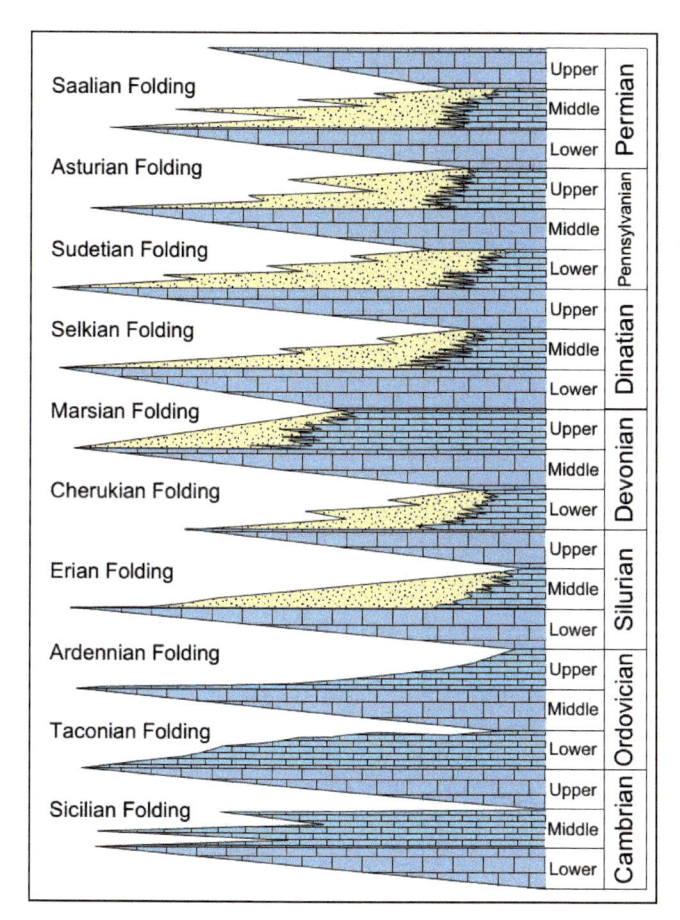

Redrafted version of Grabau's 1936 expression of Paleozoic transgressions and regressions — "pulsation theory". Note the similarity to modern-day chronostratigraphic charts showing diachronous facies and significant time gaps.

American literature, with her book, *The Promised Land* (1912), being a bestseller promoting the unprecedented opportunities for immigrants in America.

At the same time, Grabau was authoring important publications of his own. The two-volume *North American Index of Fossils* appeared in 1909 and 1910; and in 1913, *Principles of Stratigraphy* was published—one of the classics of 20th century geology—showing Grabau's instinct for synthesis that was to reach a zenith during his years in China.

Fluent in both English and German, Grabau had been in contact with geologist Johannes Walther of Germany, whose concepts of diachronous facies greatly influenced him. Grabau also recognised the need for a better classification of sedimentary rocks, separating grain size from composition. Accordingly, he introduced the terms *rudite*, *arenite* and *lutite* to distinguish different grain sizes, which could be given prefixes to describe composition, e.g. calcarenite, to describe a limestone with sand-sized grains.

Unfortunately, the First World War was to bring about a change in fortunes. Like many Americans of German descent, Grabau sympathised with the position of the Kaiser. This brought him into conflict with both his wife and colleagues at Columbia University. The situation at Columbia was further exacerbated by political infighting in the Geology department. No matter the quality of his publications, Grabau was seen as a troublemaker and by the end of the war; he had lost his university position and separated from his wife.

From adversity often springs opportunity. For Grabau, it came in the form of an offer to take on the position of Professor of Paleontology at Peking University and Chief Paleontologist to the Chinese Geological Survey. By 1920, at the age of 50, he was on his way to China to embark on an entirely new phase of his career and personal life. Grabau had just completed *A Textbook of Geology*, which provided the ideal basis for his new courses (taught in English).

The opportunity to teach in China came about because of the efforts of Deng Wenjiang, also known as V.K. Ting. Ting, as Director of the fledgling Chinese Geological Survey, recognised the need to raise the status of geology in China, both to secure its economic fortunes and because the geology of China could offer so much to research. Ting had been trained in mineralogy and physical geology at the University of Glasgow and wanted to expedite the infusion of Western science into China. Accordingly, he asked David White, the Chief Geologist of the U.S. Geological Survey, for a recommendation. It was White who pointed him towards Grabau, a very fortunate choice.

The first few years in China were exciting times for Grabau. He quickly came to grips with the geology of the country, publishing a string of papers and books, culminating in *The Stratigraphy of China*. He taught ever larger classes of students and developed a pool of Chinese researchers, who went on to lead research teams of their own, so that very rapidly, Chinese geoscience research and teaching was on a sure footing.

It was not long after Grabau had arrived in China that fragments of humanoid fossil skeletons were discovered that became known as "Peking Man" (a form of *Homo erectus*). This discovery put China on the map scientifically, with Peking becoming a centre of gravity for researchers and adventurers, alike. Grabau found himself in the midst of inspiring expatriates, including the explorer Sven Hedin, the adventurer Roy Chapman Andrews (the alleged role model for Indiana Jones and discoverer of dinosaur eggs in Mongolia) and the priest/paleontologist/philosopher Pierre Theilhard de Chardin. Working in this fertile atmosphere, Grabau's research efforts intensified.

As Grabau contemplated the stratigraphy of China and compared it with his knowledge of other parts of the world, similarities became apparent to him. Throughout the Paleozoic, there seemed to be widespread synchronous unconformities and major synchronous transgressions. This led Grabau to develop his "Pulsation Theory," essentially claiming that the Earth had been subject to episodic global sea-level change throughout its history. Recognition of transgression and regression was greatly aided by an understanding of facies diachronism—an idea he had imported from his dialogue with Johannes Walther.

Eustasy was first termed by Eduard Suess in 1888 and the concept was much discussed by T.C. Chamberlin, amongst others, in the early part of the 20th century. Unlike these earlier workers, Grabau not only discussed how to recognise sea-level change in the rock record, but presented a biostratigraphically calibrated global database that could support the synchronicity of events between locations. He also related his observations to an explanatory model (heating and expansion, then cooling and contraction of the ocean floor). It seems that Grabau arrived at the notion of eustatic sea-level change from a gradual gathering and synthesis of data, rather than simply setting out to prove the theory. Sequence stratigraphers today continue to build on the foundations he laid.

Grabau first presented his ideas at the 1933 International Geological Congress in Washington, D.C. (his only return to the country of his birth after moving to China). He subsequently published four volumes of *Palaeozoic Formations in the Light of the Pulsation Theory*, followed by his grand synthesis, *The Rhythm of the Ages*, in 1940. At the heart of this work was the detailing of regular Phanerozoic transgressions and regressions. But it was also all-encompassing, dealing with everything from the origin of life on Earth to recent ice ages. Most importantly, it provided 18 colour paleogeographic reconstructions of how the Earth had appeared at different episodes in the geological past. Given that Alfred Wegener's ideas on continental drift were still opposed by many geologists, this was a radical step. Although initially an opponent himself, Grabau used facies and fossil distribution to suppose, for example, that Greenland, north-western Europe and north-eastern America had been much closer together during the Silurian. During the Ordovician and Silurian, polar ice caps were shown over large parts of Arabia and North Africa, proof of which was not found until the 1970s.

What is even more remarkable is that this grand synthesis was prepared at times of extreme difficulty for Grabau. His health had deteriorated, with arthritis restricting his movements. At the same time, Peking was occupied by Japanese troops following the outbreak of hostilities between China and Japan in 1937. As things worsened in China, with detention once America declared war on Japan in 1941, Grabau's health deteriorated even further. Although he survived the war, he died in 1946 after a short battle with cancer.

Richard Fortey, in *Trilobite! Eyewitness to Evolution*, noted that reverence for wisdom endures in China. *"I was taken to see the grave of Professor Grabau,"* he writes. *"A Western paleontologist, who almost singlehandedly introduced modern geological principles into China in the early years of the twentieth century. He was, so I was told, 'a great teacher.' The Chinese compliment is such an important one that it is represented by a special ideogram. It was a simple grave, but obviously and lovingly tended."*

Amadeus Grabau is not a name that is known to every modern geologist, but deserves to be for his foresight, mastery of synthesis and contribution to establishing geology in the world's most populous country, China.

REFERENCES

This essay has drawn upon information from the following sources:

Johnson, M.E. 1992. A.W. Grabau's embryonic sequence stratigraphy and eustatic curve. In: Dott, R.H. Jr. (ed.) *Eustasy: The Historical Ups and Downs of a Major Geological Concept*. Geological Society of America Memoir, 180, 43-54.

Mazur, A. 2004. *A Romance in Natural History*. Garret, Syracuse, New York. 484pp.

Oldroyd, D.R. 1996. *Thinking About the Earth*. The Athlone Press, 410pp.

Pemberton, S. G., Bhattacharya, J. P., MacEachern, J. A., & Pemberton, E. A. 2016. Unsung Pioneers of Sequence Stratigraphy: Eliot Blackwelder, Joseph Barrell, Amadeus Grabau, John Rich and Harry Wheeler. *Stratigraphy*, 13, 223-243.

William Jocelyn Arkell in 1947. Photographed by Walter Stoneman and reproduced with permission under licence from The Royal Society.

William Jocelyn Arkell

It is very doubtful if today any geologist would attempt single-handedly to write a book entitled *"Jurassic Geology of the World"*, but in 1956 just such a book appeared authored by William Jocelyn Arkell. Arkell was one of the great stratigraphic geologists and paleontologists of the 20th century, although his scientific contributions encompassed a variety of geological subjects. He was a prolific author. A list of his publications includes around 150 articles and several important books, memoirs and explanatory notes to geological maps. This is all the more remarkable given that he suffered ill health for much of his life and during his career he mostly occupied unsalaried research positions.

He was born in June 1904 in Wiltshire in southern England. His father was a partner in the local brewing company (Arkell's – still in existence) and income from the brewery was to fund William's lifelong geological research activities. In 1922 he entered at New College, Oxford University and obtained a 1st Class Honours degree in Geology in 1925. Family holidays spent at Swanage on the Dorset coast had infected him with an enthusiasm for geology and for Jurassic rocks and fossils in particular. This quickly became a research theme for him as he stayed on at Oxford to obtain his doctorate. The Oxfordian ('Corallian') succession around Oxford and south into Wiltshire and Dorset was his initial research topic but he quickly expanded his interests to include the entire Jurassic of the UK. This culminated in the publication in 1933 of his superb synthesis *"The Jurassic System in Great Britain"*. This 681 page work with copious illustrations was a remarkable achievement for a young man still in his twenties and remains a valuable reference work today.

Coherent synthesis is a skill often undervalued in scientific writing, but was one at which Arkell excelled culminating in *Jurassic Geology of the World*. In *The Jurassic System in Great Britain*, building on the work of William Smith, Alcide d'Orbigny and the other early pioneers of stratigraphic classification and correlation, he not only described the British Jurassic stratigraphic units and their characteristic fossil content, but also reviewed the importance of biozones and their definition, a topic he was to return to a number of times in his career. He also linked tectonics to stratigraphy and their influence on palaeogeography. This was truly a modern synthesis and one which established his international reputation.

In the 1930s Arkell continued to be based at New College, Oxford with minimal teaching and administration duties (not unreasonable given he received no salary!) allowing maximum time for research. This resulted in a stream of important publications, including a growing interest in Jurassic ammonites, especially their taxonomy, phylogeny and value in correlation. At the same time, he was interested in the

Specimens of *Dactylioceras* from Germany. Jurassic ammonites were a crtical part of Arkell's research.

tectonic history of southern England (he presented a paper on this to the International Geological Congress in Washington in 1933) and in collaboration with the Geological Survey, contributed to the mapping of parts of Dorset, Wiltshire, Berkshire and Oxfordshire and authoring the supportive descriptive memoirs. The memoir of the geology of Weymouth and Purbeck district remains a classic and is adorned by Arkell's own skilful sketches. Much of his writing at this time was carried out during the summer months in the company of his family at their holiday chalet "Faraways" at Ringstead on the Dorset coast close to classic Corallian and Kimmeridge Clay outcrops.

The Second World War led Arkell to become a temporary civil servant assigned to the Ministry of Shipping in London. His work was cut short in 1943 when he became seriously ill, spending five months in hospital with a near-fatal chest infection. His recovery took some time limiting his ability to carry out field work until the late 1940s. Despite his enforced convalescence he continued with his publications including two major books published in 1947: *The Geology of Oxford* and *Oxford Stone*. The latter reflected his interests in building stone and their correct use in the renovation of many of historic Oxford colleges. An interesting minor publication of this time is the pamphlet *Geology and Prehistory from the Train, Oxford – Paddington*. This begins with the delightful opening passage: *"During the years 1941-1943, owing to 'circumstances arising out of the war' it was my lot to make the double journey between Oxford and Paddington and back about 100 times. The following notes are a humble offering to my fellow-travellers; compiled in the belief that to increase sources of interest is proportionally to diminish boredom."* Needless to say this work is extremely informative and written in a fluent prose that characterised all of his writing.

In 1947 Arkell was offered the opportunity to take up a research post at Trinity College, Cambridge with a room in the famous Sedgwick Museum. Here he completed a number of monographs on British Jurassic ammonites and began work on his contribution on Jurassic ammonites for the monumental *Treatise on Invertebrate Palaeontology*. As such he was now receiving specimens of Jurassic ammonites from around the world and was in correspondence with Jurassic specialists from every corner of the globe. Oil companies sent him

The Oxfordian "Corallian" succession exposed between Osmington Mills and Ringstead Bay on the British Dorset Coast. The stratigraphy of this succession was of particular interest to Arkell and close to his summer holiday home and location for much of his writing "Faraways".

considerable material including the Arabian American Oil Company (now Saudi Aramco) who at that time were getting to grips with the economically important Jurassic stratigraphy of Saudi Arabia. At their invitation, Arkell was able to visit that country and also made visits to Egypt, Algeria and Tunisia in the early 1950s.

His travel and correspondence in the post-war years placed him in an ideal position to undertake his monumental synthesis, the *Jurassic Geology of the World*. This is both a critical review of an extensive and potentially bewildering publis framework.

In the early autumn of 1956 Arkell suffered a severe stroke that left him partially paralysed. He continued with his research, but in April 1958 he passed away within a few hours of a second stroke at the young age of 53. With his passing a major contributor to stratigraphic geology was lost. His many and varied publications had seen him, amongst many other honours, elected to the Royal Society in 1947 and receive the Lyell Medal of the Geological Society of London in 1949. A shy, reserved man, his legacy was to link excellent factual descriptions to regional synthesis using well considered biostratigraphy as a framework, a process that remains as valid today as it did in the mid-20th century.

REFERENCES

This essay has drawn upon information from the following sources:

Cox, L.R. 1958. William Jocelyn Arkell 1904 – 1958. Biographical *Memoirs of Fellows of the Royal Society*, 4, 1-14.

https://www.ogg.rocks/william-j-arkell

Marie Tharp

The impetus for the plate tectonics paradigm came in part from the discovery of the spectacular geomorphology of the ocean floor from the 1940s onwards. At the center of these discoveries was Marie Tharp, a pioneering oceanographer and geologist, who was largely responsible for generating the first global map of the ocean floor and noting the rifted nature of mid-ocean ridges.

The geological map of England, Wales and part of Scotland, published by William Smith in 1815, has been called "the map that changed the world." But there is a more recent map that can rightly make the same claim. Marie Tharp, in collaboration with her career-long colleague Bruce Heezen, created the World Ocean Floor Map published in 1977. For the first time, in bright colour and a three-dimensional rendition, the true scale of sea-floor topography was clearly illustrated. An inspiration to scientists and armchair travelers, alike! The map was the product of thirty years of research led by Tharp and Heezen, including discoveries that would provide critical impetus to the development of the theory of plate tectonics.

Marie Tharp was born in Ypsilanti, Michigan in 1920. Her father was a soil surveyor for the US Department of Agriculture and this led her to move around the country, such that she attended 24 different schools before attending college. Going into the field with her father doubtlessly inspired an interest in maps, although it would take some time before this found an application in her career.

An avid reader, Tharp chose for her first degree a BA in English at Ohio University. Whilst there, she decided to take a course in geology out of general interest. This led her to review her career options and in 1944, she embarked on a Master's degree in geology at the University of Michigan. The programme was designed to attract women into the sciences by guaranteeing them jobs in the oil and gas industry. Thus, the ten students in her class were called the "PG girls" (petroleum geology girls).

After a brief spell as a junior geologist with the US Geological Survey, Tharp went to work for Standolind Oil and Gas in Tulsa, Oklahoma. Unfortunately, she found the office work unrewarding and decided to undertake a further degree in mathematics at Tulsa University, graduating in 1948. She then relocated to New York in search of a new career opportunity.

During the Second World War, the US Navy and its allies had collected a vast amount of bathymetric data from the world's oceans from numerous sonar surveys. After the war, this data became available to the scientific community to study and augment with new data from research cruises. Analysis of the data was conducted by the Lamont Geological Observatory of Columbia University (now the Lamont-Doherty Earth Observatory) in New York, led by Maurice Ewing.

Tharp arrived at Columbia in 1948 as a drafting and computational assistant to Ewing's graduate students. One of these students, Bruce Heezen, had begun collecting Atlantic ocean-floor data in 1947. Since he was often away at sea, she soon took on the responsibility for collating, organizing and eventually mapping data that he obtained. The two maintained a professional collaboration until his death (aboard a submersible, whilst collecting data off Iceland) in 1977.

Marie Tharp working on a map of the Atlantic Ocean floor in the early 1950's. Note the profiles that were to reveal the rift within the mid-ocean ridge.

Marie Tharp and Bruce Heezen with an early version of their Atlantic Ocean floor map.

Ocean-floor contour maps were classified as confidential until 1961, so Tharp and Heezen used "physiographic mapping," developed by the Columbia University geomorphologist Armin Lobeck, to portray a somewhat three-dimensional representation of ocean-floor topography. Making such maps required a combination of geological and geomorphological knowledge, plus mathematical skills to convert the raw bathymetric data into profiles, maps and diagrams. Tharp's diverse education, therefore, stood her in good stead. The process of creating profiles and then stitching these together to create the three-dimensional physiographic diagrams was a painstaking process and involved some degree of informed speculation on the geomorphology between data points. Nonetheless, the results were striking and once the first map was published in 1957 (an article on assisting in the placement of submarine cables), there was considerable interest in their production.

By the mid-1960s, the National Geographic Society had interest in the maps reaching a wider audience. This resulted in a classic series of maps being published between 1967 and 1971 in *National Geographic Magazine* that documented the ocean-floor of the Indian, Atlantic, Pacific and Arctic oceans. The Society hired an Austrian artist, Heinrich Berann, to colour the black and white originals produced by Tharp and Heezen, producing extremely striking images. Their last project, completed in 1977, was the much-celebrated map of the entire global ocean-floor, funded by the US Office of Naval Research.

Ever since the *Challenger* expedition of 1872, it had been known that the floor of the central Atlantic Ocean possessed an undersea rise; but until the detailed wartime bathymetry became available, its size and character were unknown. Soon, submerged mountain chains were also known from the Indian and Pacific oceans as the true nature of the ocean floor finally became apparent.

By 1960, the notion of seafloor spreading was beginning to take hold in the geology/oceanography community. The Princeton professor and former naval officer Harry Hess speculated, as had Arthur Holmes many years before, that convective overturn within the mantle was driving migration of the crust. Central to this argument were the discoveries regarding the nature of mid-ocean ridges. It was Marie Tharp, who in 1952 had demonstrated the existence of medial graben along ridge crests, and then with an initially skeptical Bruce Heezen, had suggested that the crust was extending at right angles to the trend of the ridges. This was consistent with the notion that the ridges were the sites of upwelling convective currents and the ocean crust migrated laterally away from the ridges at rates approximating 1cm/year. Maurice Hill and John Swallow of the University of Cambridge independently

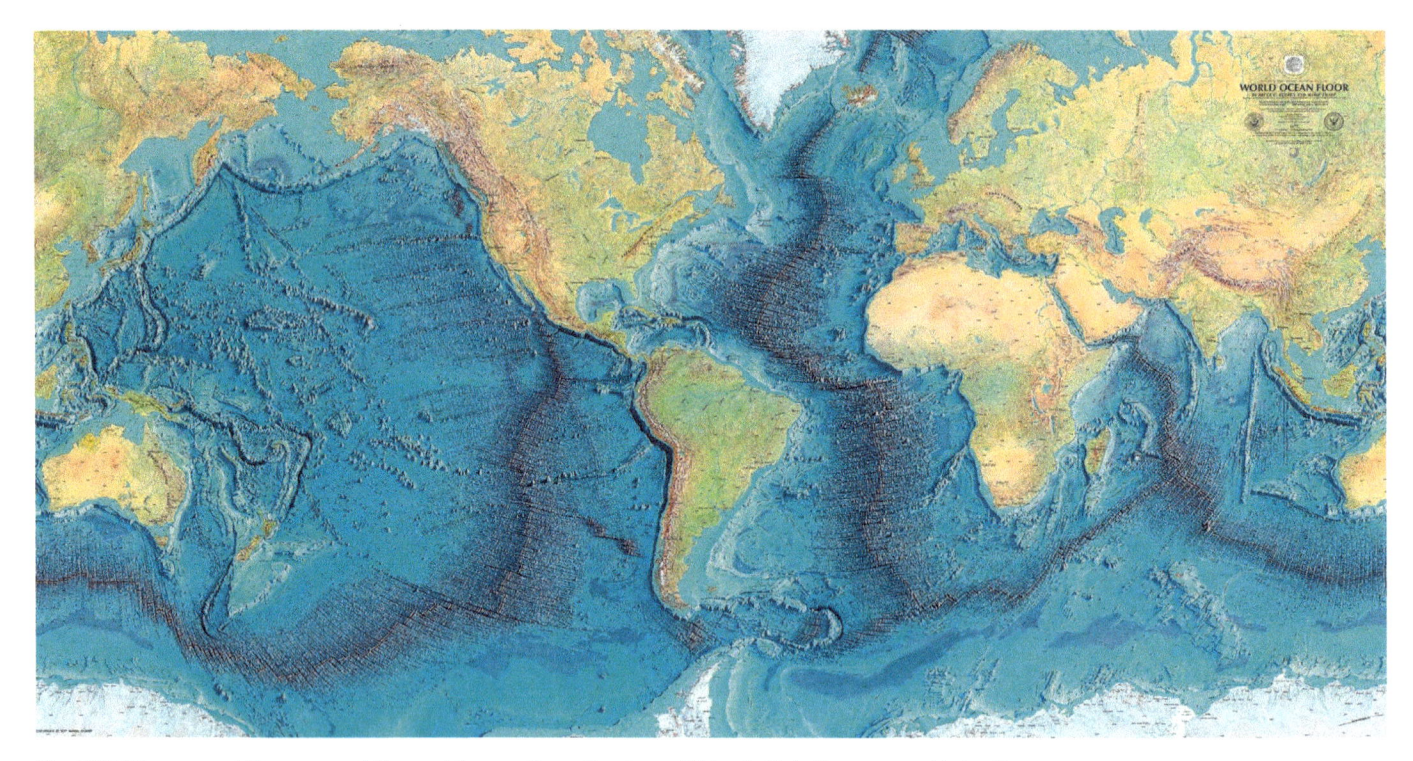

The 1977 Heezen and Tharp map of the world ocean floor. Courtesy of Woods Hole Oceanographic Institute.

observed the same median ridge valley in 1953 during a British Atlantic survey.

The rift within the Mid-Atlantic ridge was compared by Heezen and Tharp with continental rifts to ensure geomorphological similarity. Rifts were then sought on other ocean ridges. Having noticed that the location of shallow focus earthquake epicenters matched the ridge rift location in the Atlantic, Tharp and Heezen, with the support of Ewing, used global earthquake data to guide their search for ridge rifts in other oceans. The linkage between the two phenomena was strong evidence of crustal extension.

Despite her contributions that led global geoscience closer to the paradigm of plate tectonics, Tharp seems to have had little interest in this theory, content to provide the description of the ocean floor for others to interpret as they wished. For a long time, her collaborator Bruce Heezen favoured the idea of an expanding Earth, perhaps partly motivated by an intense rivalry with Harry Hess. It was their informative maps and diagrams of the ocean floor that propelled other geoscientists towards the notion of plate tectonics. For example, the illustration of offset fracture zones led Tuzo Wilson to describe transform faults, a key component in the explanation of plate movement.

During the 1950s and 1960s, female research assistants were rarely encouraged in their careers and often received little recognition for their work. Despite these prevailing attitudes, Tharp's career rapidly progressed to research geologist, research scientist and ultimately research associate at Lamont. Even so, she was not allowed to go to sea on a research cruise until 1965 and then only aboard a Duke University vessel! On the other hand, her name appears on many Lamont publications and she co-authored an important book, *Floors of the Ocean*, with Heezen and Ewing.

Thanks to her research, critical groundwork was laid for the development of the plate tectonics paradigm. This included identification of a rifted mid-Atlantic ridge, the extension of mid-oceanic ridges around the planet, and the angled nature

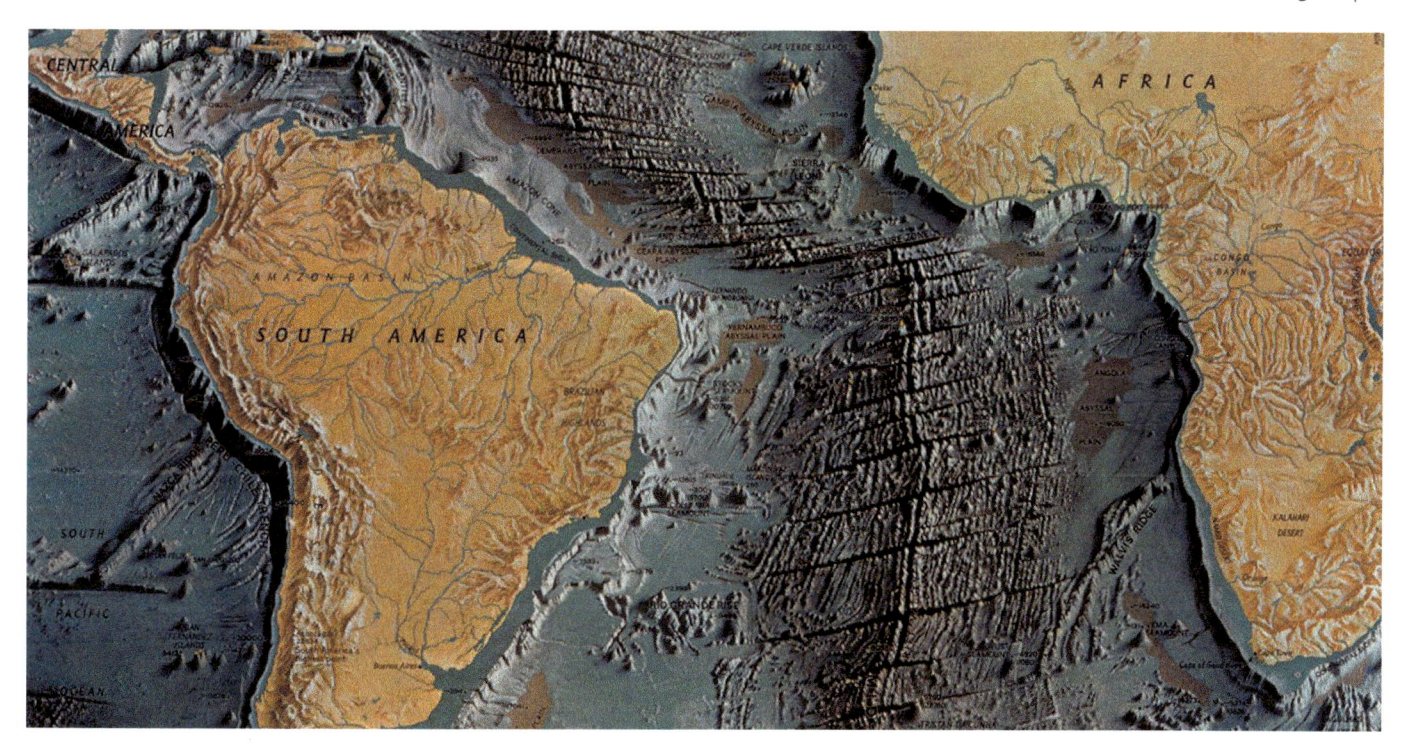

Detail of the Mid-Atlantic Ridge (adapted from National Geographic Magazine, 1968).

of faults intersecting the Mid-Atlantic ridge. Without this early knowledge, the Earth science revolution of the 1960s would have been delayed.

The magnitude of her accomplishment is perhaps best conveyed by Tharp's own words. These are her thoughts from a biographical piece she wrote upon winning the Woods Hole Oceanographic Institution's Mary Sears Woman Pioneer in Oceanography Award in 1999: "*Not too many people can say this about their lives: The whole world was spread out before me (or at least, the 70 percent of it covered by oceans). I had a blank canvas to fill with extraordinary possibilities, a fascinating jigsaw puzzle to piece together: mapping the world's vast hidden seafloor. It was a once-in-a-lifetime—a once-in-the-history-of-the-world—opportunity for anyone, but especially for a woman in the 1940s. The nature of the times, the state of the science, and events large and small, logical and illogical, combined to make it all happen.*"

REFERENCES

This essay has drawn on information within the following sources:

Barton, C. 2002. Marie Tharp, oceanographic cartographer, and her contributions to the revolution in the Earth sciences. In: Oldroyd, D.R. (ed.) *The Earth Inside and Out: Some Major Contributions to Geology in the Twentieth Century.* Geological Society Special Publications, 192, 215-228.

Felt, H. 2012. *Soundings.* Henry Holt and Company, 340pp.

Frankel, H.R. 2012. *The Continental Drift Controversy. Volume III: Introduction of Seafloor Spreading.* Cambridge University Press, 476pp.

Lawrence, D.M. 2002. *Upheaval from the Abyss.* Rutgers University Press, 284pp.

Oreskes, N. 1999. *The Rejection of Continental Drift.* Oxford University Press, 420pp.

Powell, J.L. 2015. *Four Revolutions in the Earth Sciences.* Columbia University Press, 367pp.

Tharp, M. 1982. Mapping the ocean floor – 1947 to 1977. In: Scrutton, R.A. & Talwani, M. (eds.) The Ocean Floor, John Wiley & Sons, 19-31.

Tharp, M. & Frankel, H. 1986. Mappers of the Deep. Natural History, October 1986, 1-6.

Ziad Beydoun

The Arabian Plate covers an area of almost 4.5 MM sq km and encompasses very diverse geology. Given this size and diversity, not to mention the difficulties of travel through mountainous or desert terrain, it is not surprising that the early pioneers of Middle Eastern geology focused on the outcrop mapping of relatively small areas within political boundaries. George Lees, Max Steinke, Richard Bramkamp, Rene Wetzel, Harold Dunnington, Mike Morton, Brock Powers and Ken Glennie, amongst others, published key descriptive works, the quality of which are remarkable, even by modern standards, and placed the stratigraphic description of the region on a firm footing. However, it was not until 1988 that the first true synthesis of all Middle Eastern geology was published — *The Middle East: Regional Geology and Petroleum Resource*. The author of this important book was Ziad Beydoun, a towering figure in Middle Eastern geology in the late 20th century. Beydoun's skills encompassed field geology, subsurface interpretation, integration and perhaps above all, great diplomacy, essential to synthesize the geology of the region.

Beydoun was born in Beirut in 1924, the eldest son of a district governor in Palestine. His family had a long history of political service in the Ottoman Empire. Therefore, it was not surprising that after a school education in Jerusalem and Haifa, he earned a degree in political science and history at the American University of Beirut (AUB). However, an interest in geology must have been sparked in him, perhaps by the spectacular scenery of the Lebanon and Anti-Lebanon Mountains, and he subsequently undertook a degree in geology at University of Oxford, attached to St Peter's College, graduating in 1946.

The post-war years of the late 1940s and 1950s were exciting times to be a geologist in the Middle East. Oil companies sought to build on the extraordinary discoveries of the previous decades in the Zagros Mountains, and in Saudi Arabia and neighbouring countries. In 1948, Beydoun joined the Iraq Petroleum Company (IPC). With associate companies exploring many other parts of the Arabian Plate, a geologist working for IPC could expect to be posted almost anywhere across the Middle East. Beydoun seized the opportunity and, between 1948 and 1953, he worked in Syria, Iraq, Qatar and the Trucial Coast (now the United Arab Emirates). Beydoun had a gift for languages, speaking fluent Arabic, English, French and Turkish, which must have been of great value as he moved around the region.

Ziad Beydoun, photographed circa 1960.

Wadi Doan in the Hadhramaut region of Yemen.

In 1953, he started work in south-western Arabia, now Yemen, and parts of Oman. His work in this region was to make his name in Middle Eastern geology and involved field work in some of the most remote and at that time hostile locations. It began with a one-man survey of the island of Socotra, off the Horn of Africa in the Arabian Sea, and was followed by surveys in the Hadhramaut region. The geology here is impressive, but the stratigraphy was difficult to unravel and required several field seasons of effort in order to reach the first modern description of the geology of the region. Although IPC abandoned the idea of petroleum exploration in south-western Arabia in 1961, Beydoun was able to publish the results of his field work in a seminal monograph: *The Stratigraphy and Structure of the Eastern Aden Protectorate*. This work also formed the basis for his doctorate, obtained from the University of Oxford.

In 1963, he returned to Lebanon to take up the post of assistant professor at AUB, but in 1966 returned to the industry leading Marathon Oil's evaluations of Middle Eastern geology. Beydoun enjoyed both the academic life of teaching students and writing papers, and the practical work of being an industry geologist. From 1970, he was able to enjoy the best of both worlds, as both a professor at AUB and an advisor to Marathon.

With the onset of civil war in Lebanon in 1975, Beydoun had many difficulties to overcome. Despite the horrors around him, he continued to lecture and guide students in the Geology Department of AUB. His teaching and fortitude in the face of adversity inspired many students, several of whom went on to hold key positions in the oil industry around the Middle East, providing Beydoun with a network of contacts to facilitate his regional research.

His knowledge of Middle Eastern geology was encyclopedic, possessing a superb memory of the results from wells drilled decades before, as well as the rocks he had personally encountered in his many years of field work. This placed Beydoun in the enviable position of conducting the first regional synthesis of Middle Eastern geology, published in 1988 and updated in 1991, with an emphasis on the importance of plate tectonics. Both books remain essential reading for geologists of the Middle East today.

Ziad Beydoun in the field in the Hadhramaut region in 1953.

As a result of his substantial scientific output, Beydoun was awarded the William Smith Medal from the Geological Society and the Medal of the National Order of the Cedar by the Government of Lebanon. He died in 1998, surrounded by papers and books as he worked to complete a new synthesis of the geology of Yemen. This key publication was published posthumously and continues to be an important reference work. A long-time colleague at AUB, Chris Walley, described Beydoun as "a geologist and gentleman" and Ziad (or "Don," as he was called by his close friends) will live long in the memory of all those who were privileged to meet him.

REFERENCES

This essay has drawn on information within the following sources:

http://al-bab.com/albab-orig/albab/bys/obits/beydoun.htm

http://www.britishmuseum.org/research/search_the_collection_database/term_details.aspx?bioId=92741

http://almashriq.hiof.no/ddc/projects/geology/beydoun/

http://b-ys.org.uk/journal/obituaries/beydoun-ziad

http://archives.datapages.com/data/bull_memorials/82/082010/pdfs/1876.htm

Harry Hess

Harry Hess whilst serving as an officer aboard the USS *Cape Johnson* in World War II.

In February 1945, the assault transport ship USS Cape Johnson was engaged in supporting the American troop landings at Iwo Jima, one of the key actions in the Pacific Ocean theatre of World War II. This was one of many operations in which the ship was involved, as it transported troops between bases in the Pacific, and to various battlefronts. Its military missions also helped serve a scientific purpose. The ship was equipped with sonar that allowed bathymetric surveys of the ocean to be undertaken.

The significance of this opportunity was not lost on the ship's executive officer (subsequently, its commander), Harry Hess. Hess was an academic geologist based at Princeton University, who as a U.S. Navy reservist had reported for active duty the day after the Pearl Harbour attack by the Japanese air force. Hess already had a particular interest in the geology of the oceans and understood that these new data could throw light on how the oceans formed. In fact they proved inspirational him in developing one of the most important strands of the plate tectonics paradigm — sea-floor spreading.

Harry Hess was born in New York in 1906 and attended Ashbury Park High School in New Jersey. He entered Yale University in 1923, with the intention to study electrical engineering. However, he found this subject boring and, looking for something that would give freer rein to his imagination, decided to study geology. Since he was one of only two undergraduates at the time, he took graduate as well as undergraduate courses. This was hard work, but he graduated in 1927, with the first B.Sc. in Geology at Yale.

After graduation, Hess spent eighteen months as a mineral exploration geologist in Northern Rhodesia (now, Zambia). He remarked, *"At seventeen miles a day, I developed leg muscles, a philosophical attitude toward life, and a profound respect for fieldwork."*. However, he was not destined to be an industrial geologist. He returned to the U.S. and was accepted for a PhD program at Princeton, studying the serpentinisation of a large peridotite intrusion at Schuyler in Virginia. He received his PhD degree in 1932.

Hess taught at Rutgers University between 1932 and 1933 and spent time at the Geophysical Laboratory of the Carnegie Institute in Washington, D.C. before accepting a position in the Geology department at Princeton in 1934. This would continue to be his academic base for the rest of his career.

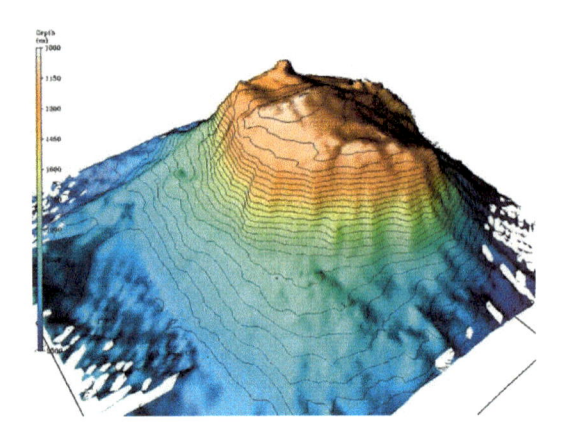

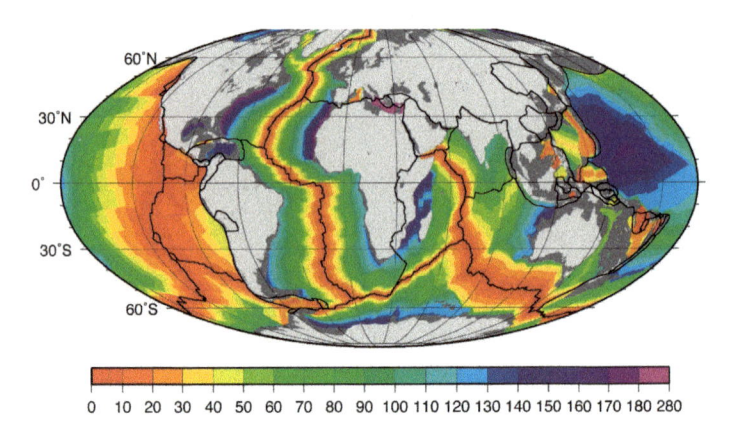

The Bear Seamount, a guyot from the North Atlantic.

The age of the ocean crust demonstrates sea-floor spreading.

Whilst undertaking his PhD research, Hess had the opportunity to participate in gravity and bathymetric surveys in the Caribbean Sea aboard the submarine USS S-48. This experience and a subsequent survey aboard the USS Barracuda in 1937 would be the catalysts sparking his interest in marine geology. The S-48 surveys were carried out along with the Dutch geophysicist, Vening Meinesz, who became a mentor to Hess as they discussed the origin of the oceanic and gravimetric features they had observed.

Of particular interest was the coincidence of negative gravity anomalies and ocean deeps adjacent to island arcs, in features Hess and Meinesz described as "tectogenes," which were ascribed to down-buckling of the crust. No explanation for the warping of the crust was given, although the concept would subsequently be adapted thirty years later as part of the process of subduction. Hess speculated that sediment infilling the basin above a tectogene would eventually be deformed as tectogenesis continued, forming mountain belts, such as the Alps. The ultramafic serpentines of his PhD research had their place in this theory, too. They were intruded from the mantle in an early stage of tectogenesis and subsequently uplifted and deformed. Hence, serpentine belts, which Hess knew from the Alps and many other ancient mountain belts, were evidence for deformed and uplifted tectogenes. These ideas now seem far from plate tectonics but, nonetheless, were stepping stones towards that understanding.

Hess's research activities were put on hold in 1941, with the entry of the U.S. into World War II; although, as already mentioned, the sonar aboard the USS Cape Johnson allowed him to gather valuable bathymetric data from the Pacific Ocean. After the war, he was able to evaluate the implications of these data. A surprising discovery was the presence of twenty deeply submerged, reefless, flat-topped seamounts that he named "guyots" after Arnold Guyot, the Swiss oceanographer, who founded Earth Sciences at Princeton. A further 140 guyots were identified through examination of bathymetric charts from the U.S. Navy's Hydrographic Office. They were circular or oval in shape and ranged from two to sixty miles in diameter. Their origin presented Hess with a puzzle. The flat tops suggested erosion by wave action, but if this were so, why were reefs not present?

Hess's first solution was to suggest that these guyots were Precambrian oceanic islands formed before the advent of reefal organisms. During the time since their formation, the sea level had risen because of constant sedimentation on the ocean floor, along with isostatic adjustments. His observations and ideas on guyots were published in 1946, in a paper that firmly established Hess's already considerable scientific reputation, even though the erroneous theory on the origin of guyots would subsequently be rejected, not least by Hess, himself.

After the war, increasing amounts of data on the ocean crust, ocean floor topography and sedimentation were gathered by the international scientific community at large. Hess saw his role as the synthesiser of these data, providing explanations of what they implied, and his ideas began to evolve very rapidly. A strong believer that science moves forward in an iterative process, he was not afraid to advance a hypothesis and then overturn it as new data emerged. This constant development and refining of hypotheses were important steps towards the concept of sea-floor spreading, which he first put forth in 1960. One might say that Hess carried out his thinking in public.

Building on ideas first introduced by Arthur Holmes and others, Hess believed that mantle convection might provide an explanation for deformation of oceanic crust. By 1953, he had turned his attention to the mid-ocean ridge known to be in the Atlantic. Having previously considered this to be an old, folded mountain belt, he now related its origin to upward convection in the mantle driving intrusion and uplift. When this convection ceased, the ridge subsided. However, he was soon forced to reverse the notion once he realized that rising convection causes deserpentinisation (serpentine is transformed into olivine at temperatures above 500°C and water is released) and, hence, sinking of the ocean floor. This concept also provided him with a mechanism to sink guyots deep below sea level. (The fact that some guyots had Late Cretaceous fossils at their crests challenged Hess's original idea of them being Precambrian islands.)

The late 1950s saw Hess constantly updating his models for the origin of oceanic features, such as mid-ocean ridges, guyots and island arcs. The tectogene hypothesis became increasingly challenged. As the debate for and against continental drift grew, Hess was initially a "fixist," arguing in 1955 that Early Palaeozoic folded mountain belts in North America and north-western Europe might continue beneath the Atlantic and be hidden by younger sediments. However, by 1959, he was happy to switch to "mobilism" as the first palaeomagnetic studies demonstrating polar wander paths were published. Interaction with the Australian geologist, Warren Carey, who favoured an expanding Earth, but in doing so, promoted mobilist ideas, may also have been an influence.

By 1960, Hess was ready to assess all the oceanic features, which he had spent much of his career discussing, in terms of mobilsm. The result was a manuscript entitled Evolution of the Ocean Basins, a report to the Office of Naval Research widely circulated in 1960, but not formally published until 1962 as History of the Ocean Basins in a Geological Society of America publication. Hess envisaged that oceans grew from their centers, with molten material (basalt) rising up from the Earth's mantle along the mid-ocean ridges, driven by mantle convection. The presence of a rift valley at the center of ocean ridges, as detected from bathymetric surveys by Mary Tharp and Bruce Heezen, was a crucial observation that helped Hess to develop his ideas.

The extrusion of basaltic lava created new sea floor, which spread away from the ridge in both directions. The ocean ridge was thermally expanded and, consequently, higher than the ocean floor further away. As spreading continued, the older ocean floor cooled and subsided to the level of the abyssal plain, which is approximately 4 km deep. Ocean trenches were areas where ocean floor was destroyed and recycled — a point that Hess expanded upon in a joint paper with Robert Fischer, of the Scripps Institute of Oceanography, in 1963. His long-standing conundrum of the origin of guyots could also be explained by this theory; i.e., they evolved from wave-eroded volcanic peaks that formed at ridge crests and then, as spreading moved them away from the ridge, they subsided with the cooling oceanic crust they were resting upon.

Hess did not initially call this theory "sea-floor spreading" — that term was introduced by Robert Dietz in a 1961 Nature paper, which contained many similar ideas to those that had been circulated by Hess in 1960. Once introduced, the term became widely accepted.

Hess was well aware that his ideas were both provisional and controversial. His History of the Ocean Basins paper contained the line, "*I shall consider this paper an essay in geopoetry;*" thereby, emphasising its speculative nature. If it were poetry, it scanned well. The paper explained many observations in an integrated way and offered the solution that proponents of continental drift had long sought — a mechanism for continental movement along the conveyor belt of sea-floor spreading.

Simplification of the sea-floor spreading hypothesis advanced by Harry Hess.

Support for Hess's ideas was soon to appear when, in 1963, the British geologists, Fred Vine and Drummond Mathews, published a paper in the journal Nature, noting that there was a symmetrical pattern of magnetic stripes (positive and negative magnetic anomalies relating to magnetic pole reversals) on either side of the mid-ocean ridges. In addition, when the basalts of the sea floor were dated, they were found to be the same age at similar distances away from the ridge on each side. This suggested that the ocean floor was created at the mid-ocean ridges, and then progressively moved away from the ridge, just as Hess had speculated.

Outside of his work on marine geology, Hess was also involved in many other scientific endeavours, including the Mohole project (1957–1966), an investigation into the feasibility and techniques of deep sea drilling that eventually gave rise to the Deep Sea Drilling Programme and its successors. He continued to be involved in the U.S. Naval Reserve, rising to the rank of Rear Admiral. He was also an active adviser to U.S. governmental science programmes, including the Lunar Exploration Missions. Hess died from a heart attack in Woods Hole, Massachusetts, on August 25, 1969, while chairing a meeting of the Space Science Board of the National Academy of Sciences.

Hess was one of the truly great American geologists, who took full advantage of being at the center of the explosion of data gathering from the marine realm in the mid-20th century. He had the rare ability to overturn his ideas when new data demanded it and did not hesitate to say publically that he had been wrong. His greatest contribution to geology was the notion of sea-floor spreading, without which the development of the paradigm of plate tectonics would have been delayed. Hess said of himself in his speech accepting the Geological Society of America's Penrose Medal in 1966, *"As a geologist who has often guessed wrong, I deeply appreciate the generosity of the society in balancing my errors against deductions of mine not yet proven incorrect. I am pleased to come out with a positive balance."*

REFERENCES

This essay has drawn upon the following works:

Frankel, H.R. 2012. *The Continental Drift Controversy, Volume III Introduction of Seafloor Spreading.* Cambridge University Press, 476pp.

Lawrence, D.M. 2002. *Upheaval from the Abyss. Rutgers University Press*, 284pp.

Le Grand, H.E. 1988. *Drifting Continents and Shifting Theories.* Cambridge University Press. 313pp.

Oreskes, N. 1999. *The Rejection of Continental Drift.* Oxford University Press, 420pp.

Fred Vine

Fred Vine, photographed at Princeton University in 1968.

The development of plate tectonics involved a number of individuals, and teams of scientists, each working on separate strands of the theory. At the beginning of the 1960s, the concept of seafloor spreading was introduced by Harry Hess and by Robert Dietz, who sought to explain observations from bathymetric and gravity surveys of the World's oceans. Seafloor spreading was an intriguing hypothesis, suggesting that ocean crust rose up from the Earth's mantle along mid-ocean ridges, and then spread away from these ridges in both directions to be subducted in ocean trenches.

But how could the notion of crust moving like a conveyor belt be proven? The answer was to come from paleomagnetics, studies of the Earth's ancient magnetic field as recorded in the rock record. At the forefront of this work was Cambridge University research student, Fred Vine. In 1963 Vine, together with his supervisor Drummond (Drum) Matthews, developed a speculative proposal tying together seafloor spreading and paleomagnetism as a means of explaining the patterns of magnetic anomalies found parallel to mid-ocean ridges. They suggested that oceanic crust was not just a conveyor belt, but also a tape recorder, recording the Earth's past magnetic polarity and, in so doing, confirming seafloor spreading. It was now possible to accept the notion of large-scale lateral crustal motion, accelerating the progress towards a complete theory of plate tectonics.

PALEOMAGNETIC THEORY

The alignment and inclination of the magnetic properties of minerals in certain rocks (e.g. basalts) is effectively fossilised at the moment when the rock is formed. At the beginning of the 20th century, the French geophysicist Bernard Bruhes recognised that rocks could be classified into two groups according to their magnetic properties. One group has so-called normal polarity, characterised by the magnetic minerals in the rock having the same polarity as that of the Earth's present magnetic field. That is to say, the magnetic field is oriented south to north (why a compass needle points north). The other group, however, has reversed polarity, indicated by a polarity alignment opposite to that of today's magnetic field. That is to say, oriented north to south (such that a compass needle would point south).

Based on the study of (mostly igneous) rocks at outcrop, it progressively became known that magnetic polarity has periodically switched through the course of geological time. In 1929, the Japanese geophysicist Motonori Matuyama recognised periodic magnetic polarity switching during the Pleistocene, the first step towards a magnetic polarity timescale. It would take the progressive integration of the observations of magnetic polarity reversals in the rock record with radiometric data on their ages for the notion of a global magnetic polarity reversal record through geological time to be accepted. When Vine began his studies, this was far from a fully agreed upon concept amongst many geologists and geophysicists.

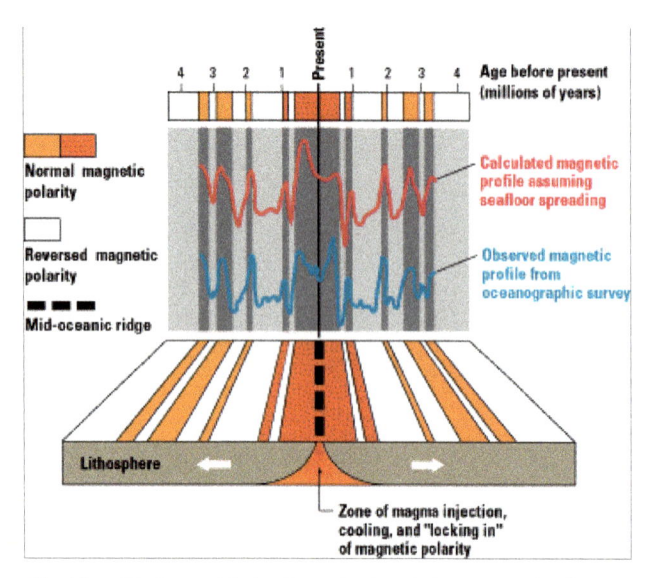

The Vine-Mathews-Morley hypothesis relating reversals of paleomagnetic polarity with seafloor spreading.

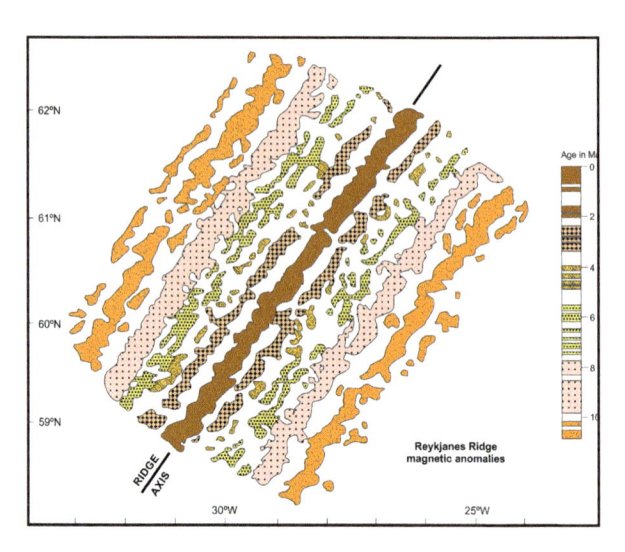

Symmetric striped pattern of magnetic anomalies on the Reykjanes segment of the Mid-Atlantic Ridge, south-west of Iceland. The positive anomalies are shaded according to their age, as indicated in the vertical column.

APPARENT POLAR WANDER PATHS

During the 1950s, studies of paleomagnetism began to present the geoscience community with some intriguing results, especially for those with an interest in the much disputed notion of continental drift. Highly sensitive magnometers were developed that allowed the direction and inclination of the remanent magnetism preserved in some rocks to be determined. A growing database of this information from rocks of different ages and from different continents was developed. This suggested that the positions of continents in relation to the poles had moved through time.

The 'apparent polar wander' was initially explained by movement of the poles through time, as opposed to movement of the continents. However, it soon became evident that each continent had a distinctive apparent polar wander path. Therefore, the continents must have moved relative to each other. This interpretation, coupled with the observation that the latitudinal positions of continents could be different from those today (as determined by magnetic inclination), aroused the suspicion of experts in marine geology, such as Harry Hess, that drifting continents were viable.

RECOGNISING PALEOMAGNETIC REVERSALS IN THE OCEANS

At the same time as paleomagnetic data were being gathered on land, magnetometers were developed to be towed behind ships that enabled magnetic surveys of the ocean floor. These studies revealed regions of magnetic anomalies, alternating strips with a magnetic intensity greater or lesser than the Earth's mean field today. These alternating stripes lay parallel to ocean ridges.

When visualised on a map of the seafloor, the pattern resembled that of the stripes on a zebra. However, what it meant was uncertain. The general assumption was that it was a local phenomenon related to compositional differences in the rocks from which the records came. Fred Vine would come to a different set of conclusions, not least because from the beginnings of his interest in geology, he was conscious of the 'big picture,' and in particular continental drift, the forerunner of plate tectonics.

VINE'S EARLY INFLUENCES

Fred Vine was born in Chiswick in west London in 1939. In April 1955, at the age of 15, when he was on his Easter holiday and studying for upcoming exams, he opened a textbook on geography and came across a diagram showing the approximate fit of the Atlantic coastlines of South America and Africa. According to Vine, "*In the text, it stated that although it had been suggested on the basis of this fit that these continents were once part of a supercontinent that subsequently split and drifted apart to form the South Atlantic Ocean, geologists had no idea whether there was any truth in this hypothesis. I was struck at once both by the boldness of the idea that seemingly stable continents might have drifted across the face of the Earth in the past, and by the fact that we did not know whether this had occurred. It seemed to me that one could hardly conduct any meaningful study of the history of the Earth until one had resolved this issue. Surely there must be some way of proving or disproving the concept of continental drift.*" It would come to pass that it would be Vine, himself, who helped supply the proof.

In 1959, Vine began his studies in natural sciences at St. John's College, Cambridge. In addition to mathematics and physics, he opted to study geology. Cambridge was a university with strength in geophysics and with some staff, like Brian Harland, positively inclined towards continental drift. In January 1962, the university hosted the 10th Inter-University Geological Congress where the guest speaker Harry Hess outlined his ideas on seafloor spreading. The lecture had a profound effect on Vine, who summarised much of Hess's work in his own talk to the student geological society, the Sedgwick Club, later that year humorously entitled "HypothHESSes". Also in 1962, he attended a lecture by the doyen of British paleomagnetists, Patrick (later Lord) Blackett, for whom continental drift was axiomatic, given the evidence from paleomagnetic data on apparent polar wander and paleolatitude.

MAGNETIC REVERSALS AND SEAFLOOR SPREADING

It is not surprising then, that at the outset of his PhD in the autumn of 1962, Vine was contemplating his research topic through the lens of continental drift. That topic was to review published magnetic surveys and traverses at sea, and the methods used in interpreting them. He began by evaluating the magnetic survey data gathered in 1962 over the Carlsberg Ridge, an oceanic ridge in the Indian Ocean. Vine has admitted he approached this work with seafloor spreading in mind. He "*was particularly looking for some record of drift and spreading.*" With spectacular insight, he rapidly concluded that the patterns in magnetisation in the vicinity of the ridge were due to alternations of reversed and normal magnetisation of the rocks forming the ridge and the ocean floor. These patterns were not local. They were the result of reversals of the Earth's magnetic field.

This theory was neatly linked to seafloor spreading. As the emergent seafloor cooled, it acquired a magnetisation in keeping with the geomagnetic field prevailing at the time. Later, as this crust was displaced by new material, if the Earth's magnetic field reversed, the newer material would be magnetised in the opposite direction to the adjacent, older crust. This led to repeated alternations and, thus, the magnetic anomaly (zebra) stripes that had been observed in marine palaeomagnetic records since the 1950s.

Vine reported his ideas to his supervisor, Matthews, who arranged for publication of a short paper in the journal Nature in 1963 — *Magnetic anomalies over ocean ridges*. Despite the importance of this paper, it took time for its conclusions and its significance to be accepted by the scientific community at large. There were a number of doubts. Firstly, the idea of episodic reversals of magnetic polarity through time was not accepted by everyone. Moreover, Vine and Matthews had combined this with the controversial notion of seafloor spreading.

TESTING THE HYPOTHESIS

Support for the Vine-Matthews hypothesis came when Harry Hess and the great Canadian geologist, Tuzo Wilson, came to Cambridge as visiting researchers in 1965. They encouraged Vine to consider that the hypothesis implied a uniform rate of spreading, which could be tested by dating the crust. It also implied that anomalies should be symmetrical on either side of an ocean ridge.

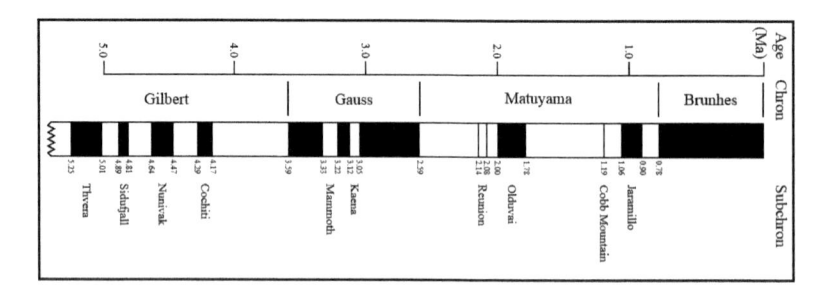

Modern understanding of Late Cenozoic paleomagnetic reversals.

Vine and Wilson collaborated to explain the magnetic anomaly patterns seen around the Juan de Fuca Ridge off Vancouver Island (*Magnetic anomalies over a young ocean ridge off Vancouver Island*, published in the journal *Science* in 1965). This ridge was, according to Wilson, bounded by transform faults, one being an extension of the San Andreas Fault. Wilson had introduced the concept of transform faults in 1965. These faults displace mid-ocean ridges and are an aspect of the mechanism by which lateral movement (spreading) of the ocean crust is accommodated.

Vine and Wilson noted the symmetry of the magnetic anomalies on either side of the ridge and, moreover, that this pattern could be recognised across the transform faults once the faults were restored to a pre-movement position. They next addressed the age of the anomalies. If spreading occurred at a uniform rate, then there should be a correlation between the widths of the normal and reversed magnetised stripes and their ages. If the ocean floor forms at 4 cm a year (2 cm on each side of the ridge), then a strip of ocean floor represents a record of 10,000 years of the Earth's magnetic history. These ages should match time scales based on continental rocks, where the age of reversals had been determined.

Vine and Wilson were able to show that a predictive model of anomaly width matched closely, but not perfectly, with the observed data. A near-perfect match was achieved after Vine, when attending the 1965 meeting of the Geological Society of America, learnt of a new short reversal event (the "Jaramillo Event") c. 0.9 Ma. The paleomagnetic timescale, when combined with the assumption of a constant rate of spreading outward from an ocean ridge, precisely predicted the symmetrical pattern of anomalies on either side of that ridge. The evidence for seafloor spreading now seemed undeniable.

Corroboration was provided by the work of Neil Opdyke and Walter Pitman at the Lamont Geological Observatory at Columbia University in New York. Pitman noted identical symmetrical patterns in paleomagnetic data collected by the research vessel Eltanin from the East Pacific Rise (the data were said to be *"too perfect"* by some who opposed seafloor spreading). Opdyke demonstrated that the continental record of magnetic reversals, including the Jaramillo Event, could be detected in analyses of sediments from the South Atlantic, with ages constrained by microfossil analysis.

In November 1966, a by-invitation-only symposium on continental drift was held in New York. Vine presented a summary of the interpretation of magnetic anomalies around mid-ocean ridges, including new data collected by an airborne magnetometer from the Reykjanes Ridge south-west of Iceland. Although the debate was vigorous (a notable exchange being - Frank Press: *"Have you tested the symmetry statistically?"* Fred Vine: *"I never touch statistics. I just deal with the facts"*), there was now little doubt that paleomagnetic evidence supported the notion of seafloor spreading and helped determine its rate. Vine ably summarised the situation in his paper, *"Spreading of the ocean floor: New evidence"* (*Science*, 1966).

IMPACT OF THE HYPOTHESIS

The stage was now set for evolution of continental drift into plate tectonics, with recognition of the segments of the Earth's crust and their motions relative to one another. That would be a story for others to pursue. As a result of his close professional interaction and friendship with Hess, Vine took a lecturing position at the University of Princeton and returned to the UK in 1970 to be based at the University of East Anglia, where he has spent the rest of his academic career, expanding his interests to ophiolites and the electrical conductivity of the crust.

It should be noted that at the same time as Vine and Matthews were developing their 1963 hypothesis, a Canadian geophysicist, Lawrence Morley, was attempting something similar, only to find his paper rejected by both Nature and the Journal of Geophysical Research ("Such speculation makes interesting talk at cocktail parties, but it is not the sort of thing that ought to be published under serious scientific aegis," wrote one withering reviewer). Morley finally had his work published in 1964, leading some to use the term 'Vine-Matthews-Morley hypothesis' for the marrying of seafloor spreading with magnetic anomaly patterns.

Paleomagnetists are often fondly called "paleomagicians" by other geoscientists. There is no doubt that without the insight provided by studies of paleomagnetism, the apparently magical movement of the continents would have not have been demystified. In this respect, Fred Vine's insightful interpretations were to prove revolutionary.

REFERENCES

This essay has drawn upon the following works:

Frankel, H.R. 2012. *The Continental Drift Controversy; Volume IV: Evolution into Plate Tectonics*. Cambridge University Press, 675pp.

Glen, W. 1982. *The Road to Jaramillo*. Stanford University Press, 459pp.

Hallam, A. 1973. *A Revolution in the Earth Sciences*. Oxford University Press, 127pp.

Kious, W.J. & Tilling, R.I. 1996. *This Dynamic Earth: The Story of Plate Tectonics*. U.S. Department of the Interior, 77pp.

Lawrence, D.M. 2002. *Upheaval from the Abyss*. Rutgers University Press. 284pp.

Le Grand, H.E. 1988. *Shifting Continents and Shifting Theories*. Cambridge University Press, 313pp.

Molnar, P. 2015. *Plate Tectonics: A Very Short Introduction*. Oxford University Press, 136pp.

Morley, L.W. 2001. The Zebra pattern. In: Oreskes, N. (ed.) *Plate Tectonics: An Insiders History of the Modern Theory of the Earth*. Westview Press, 67-85.

Oreskes, N. 2013. How plate tectonics clicked. *Nature*, 501, 27-29.

Powell, J.L. 2015. *Four Revolutions in the Earth Sciences*. Columbia University Press, 367pp.

Vine, F. J. 2001. Reversals of fortune. In: Oreskes, N. (ed.) *Plate Tectonics: An Insiders History of the Modern Theory of the Earth*. Westview Press, 46-66.

https://web.archive.org/web/20131210233929/http://sounds.bl.uk/related-content/TRANSCRIPTS/021T-C1379X0025XX-0000A0.pdf

John Tuzo Wilson

John Tuzo Wilson in 1992. Photograph by Stephen Morris.

The eminent biologist Theodosius Dobzhansky once stated, "*Nothing in biology makes sense except in the light of evolution*". Almost all geologists working today would be happy to rephrase this as "almost nothing in geology makes sense except in the light of plate tectonics". The decade from 1960 to 1970 saw a revolution in Earth sciences, with the acceptance of the general tenets of continental drift, as pioneered by Alfred Wegener several decades previously, and their modification into the single, widely accepted paradigm of plate tectonics. At the centre of this revolution was one of the greatest Canadian scientists of the 20th century — John Tuzo Wilson.

Wilson was a dynamic individual who contributed three key ideas to plate tectonic theory: mantle hot spots to explain volcanoes located away from plate boundaries, transform faults, and supercontinent cycles in which moving continents rift, collide and then break apart again. All of these ideas were developed when he was in his fifties, following a remarkable revision of his views on global tectonic processes.

Wilson was born in Ottawa in 1908. To his family he was known as Jack, or Jock, but during his professional career he began to use his middle name, Tuzo (his mother's maiden name) to avoid confusion with another J.T. Wilson. Whilst at high school, he obtained summer employment with the Geological Survey of Canada, which provided him not only with his introduction to geology, but also the skills needed to carry out fieldwork in wilderness areas.

Having developed an interest in science, he decided to study geology and physics at the University of Toronto. There was no degree in geophysics when he enrolled in 1926, and it was only through special permission that he became the first student to graduate from the university with a joint degree in physics and geology. Wilson then spent two years at the University of Cambridge before returning to North America to undertake a PhD at Princeton University. This involved mapping the Beartooth Mountains of Montana, where he carried out fieldwork on his own, including the ascent of Mount Hague, a flat-topped mountain over 3,700 m high. No doubt the geology of this region inspired an interest in mountain-building that would ultimately lead to his contributions to plate tectonic theory.

Wilson joined the Geological Survey of Canada after his PhD graduation and

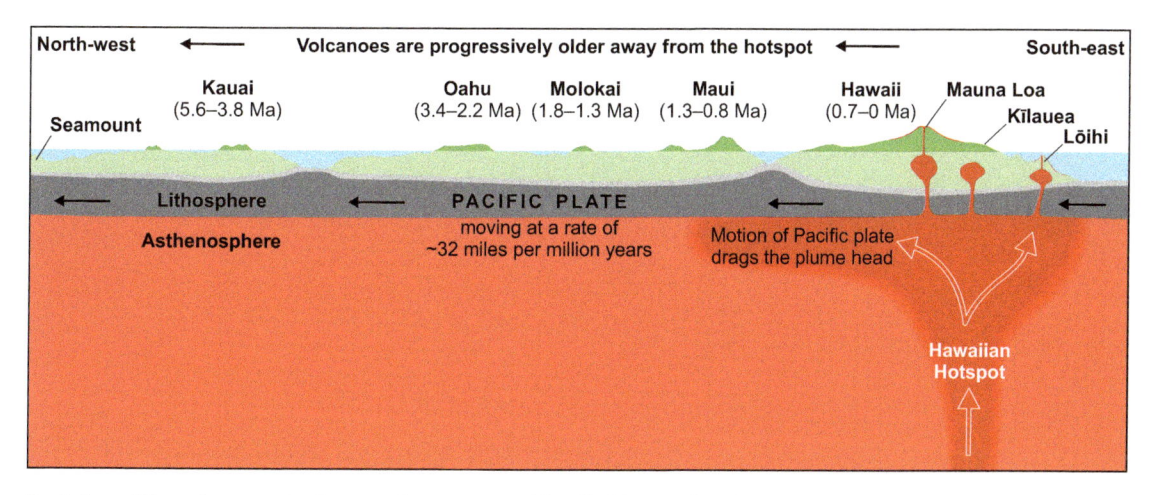

Depiction of Hawaiian mantle plume and movement of Pacific Plate as first deduced by Wilson.

carried out fieldwork in southern Nova Scotia, Quebec and the Northwest Territories. In the remote Northwest Territories, he pioneered the use of aerial photographs to delimit structures such as faults and dykes, ground-truthing where he could with fieldwork.

In 1939, the Second World War intervened and he was posted to England as a lieutenant with the Royal Canadian Engineers. His abilities impressed his senior officers and he returned to Canada at the end of the war as Colonel, Director of Army Operational Research and supervised Exercise Musk Ox, a pioneering attempt to test a convoy of new snowmobiles in severe Arctic conditions.

In 1946, it was time for a return to geoscience. He was invited, despite his relatively limited experience, to be the Professor of Geophysics in the Department of Physics at the University of Toronto, where he had studied as an undergraduate. His initial research continued, developing the understanding of the Canadian Shield that had begun with his Survey fieldwork. This included pioneering work using radiometric geochronology to understand the presence of different Archean cratons.

By 1950, Wilson was increasingly interested in global geology and mountain-building in particular. Noting that many island groups and portions of mountain chains form well-defined arcs, he collaborated with the applied mathematician Adrian Scheidegger, to show that, under certain conditions, an Earth consisting of a hot core, a rapidly cooling upper mantle, and an already cool crust, could fail in arcuate forms. Wilson was thus a "contractionist" at this time, contraction being a popular pre-plate tectonics concept to explain broad crustal structures. Before his acceptance of mobile continents in the early 1960s, he also considered the possibility of an expanding Earth.

His 1950 paper with Scheidegger brought him to the attention of the international Earth science community and initiated travel to visit overseas universities and conferences, a prominent aspect of his career thereafter. It was during the 1950s that Wilson became involved in international scientific committees including the International Union of Geodesy and Geophysics (IUGG). This group was deeply involved in International Geophysical Year, in 1957, the inaugural event of which was held in Toronto. Given the Cold War politics of the time, the event was remarkable for the involvement of scientists from the Soviet Union and China, in no small part due to the visits by Wilson to those countries.

At some point at the beginning of the 1960s Wilson accepted the idea of an Earth with moving continents and applied these ideas to his Canadian Shield research. Whilst most researchers focussed on the processes that would be known as plate tectonics observable today or in the relatively recent geological past, Wilson sought a mechanism to explain global tectonics for all geological time. By 1962 he felt able to state that "*The fact that two provinces of the Canadian Shield have been together during post-Cambrian time does not*

STAGE	EXAMPLES	DOMINANT MOTIONS	CHARACTERISTICS
Embryonic	East Africa Rifts	Uplifts	Rift valleys
Young	Red Sea, Gulf of Aden	Spreading	Narrow seas with parallel coasts and central depression
Mature	Atlantic Ocean	Spreading	Ocean basin with active mid-ocean ridges
Declining	Pacific Ocean	Shrinking	Island arcs and adjacent trenches around margins
Terminal	Mediterranean Sea	Shrinking	Young mountains and uplifts
Relic Scar (Geosuture)	Indus Line in the Himalayas	Shrinking and uplifts	Young mountains

Simplified version of the Wilson Cycle.

necessarily mean that they were formed close together or that the sediments lying on one province were derived from the province now besides it." By 1963, he was contributing actively to the development of plate tectonic theory with a remarkable series of papers that linked apparently unrelated processes such as ocean basin formation (rifting) and mountain building (convergence) into a connected dynamic model.

His first key paper relaterd to plate tectonics explained the presence of volcanoes, such as the Hawaiian Islands, far from mid-ocean ridges. Such island chains are young compared with the continents, but show a trend towards older ages with increasing distance from a mid-ocean ridge. This suggests ocean-floor spreading from these ridges, driven by mantle convection currents. But why the volcanic activity in these remote islands? Wilson proposed that the source of volcanic rock for the Hawaiian Islands is a plume rising from a 'hot spot' within the stable core of a mantle convection cell. As the Pacific lithospheric plate moves across this fixed source, older islands of the chain are carried 'downstream'. This in turn allows for the determination of velocity of plate movement relative to the hot spot.

In 1965, he followed this discovery with the idea of a new type of plate boundary, transform faults. These faults slip horizontally, connecting oceanic ridges (divergent boundaries) to ocean trenches (convergent boundaries). Transform faults were regarded as the missing piece in the puzzle of plate tectonic theory. They allow for plates to slide past each other without any oceanic crust being created or destroyed. Wilson's recognition of these features stemmed from a winter term spent at the University of Cambridge where he was able to spend time in the company of some of the other great plate tectonic pioneers: Bullard, Vine, Mathews and Hess.

Wilson then turned his attention to the life history of oceans and in particular the repeated cycles of ocean opening and closure that must have occurred through geological time. Noting that no pre-Mesozoic ocean crust now exists in situ, he recognised that ocean formation and destruction must be sought indirectly through the identification of suture zones. The Earth's oceans could be categorised in terms of their stage of maturity within a life-cycle, beginning with rifting and ending with accretion as continents converge, then rifting once again. This supercontinent cycle is now usually described as the Wilson Cycle and, although further developed by subsequent workers, remains the fundamental description of the consequence of plate tectonics.

Wilson's renunciation of his opposition to continental drift and mantle convection and his subsequent championing of plate tectonic theory is a wonderful example of the openness of mind that great scientists should possess. Before the early 1960s, proponents of continental drift had focused on the evidence for the process in the period from the Mesozoic to the Recent. Wilson's field experience was extensively with Precambrian and Paleozoic rocks, so in his opinion, any theory of global tectonics needed to be applicable throughout geological time. Once he had realised that continental drift could be applied to older rocks with, for example, the closure of a proto-Atlantic (Iapetus Ocean) in the Paleozoic, he was able to fully embrace and make key contributions to the rapidly emerging plate tectonic theory, culminating in his theory of supercontinent cycles, the unending "dance" of the continents.

Towards the end of the 1960s, Wilson began to focus his energies of the growth of the University of Toronto. He was appointed Principal of a new suburban college, Erindale College. Nonetheless, he continued to be engaged in the development of plate tectonic theory and proved himself an eloquent champion against naysayers. This included the presentation of the Canadian television series 'Planet Earth'.

On his retirement from Erindale College at the age of 65, he was given a new challenge by the Premier of Ontario - to be the Director General of the Ontario Science Centre. Under his leadership this became a hugely successful 'hands-on' museum of science.

In 1985, he retired from this post in characteristically unconventional style — entering his farewell banquet in a rickshaw pulled by a science student. This incident reflected his larger than life character and also his love for Chinese culture. Influenced by his visits to China in the late 1950s, he published a popular book about the country in 1960: *One Chinese Moon*. He also imported a Chinese junk ("*Mandarin Duck*") from Hong Kong to be used near his summer cottage at Go Home Bay on Lake Huron, north of Toronto.

Wilson had a remarkable capacity to assimilate detailed information and then arrive intuitively at simple, yet elegant models. He was a truly global geologist who travelled to almost every corner of the globe in an effort to find that "*beneath all the chaotic wealth of detail in a geological map lies an elegant, orderly simplicity.*" As with many great scientists he was willing to change his mind when faced with new evidence.

He passed away in 1993, the recipient of numerous honours including the Order of the British Empire for his military service and a Companion Order of Canada for his scientific accomplishments. Outside the Ontario Science Centre is the Continental Drift Monument. It imagines a huge fixed spike driven far into the Earth. It shows how during Wilson's lifetime the Earth's surface would have broken free from around the spike and been dragged two metres past it towards the west — an ingenious tribute to an ingenious scientist.

REFERENCES

This essay has drawn upon information from the following sources:

Frankel, H.R. 2012. *The Continental Drift Controversy. Volume IV: Evolution into Plate Tectonics*. Cambridge University Press, 675pp.

Garland, G.D. 1995. John Tuzo Wilson. 24 October 1908 – 15 April 1993. *Biographical Memoirs of Fellows of the Royal Society*, 41, 534-552.

Hoffman, P.F. 2014. Tuzo Wilson and the acceptance of pre-Mesozoic continental drift. *Canadian Journal of Earth Science*, 51, 197-207.

Lawrence, D.M. 2002. *Upheaval from the Abyss*. Rutgers University Press, 284pp.

Le Grand, H.E. 1988. *Drifting Continents and Shifting Theories*. Cambridge University Press. 313pp.

Molnar, P. 2015. *Plate Tectonics: A Very Short Introduction*. Oxford University Press, 136pp.

Oldroyd, D.R. 1996. *Thinking About the Earth*. The Athlone Press, 410pp.

Polat, A. 2014. John Tuzo Wilson: a Canadian who revolutionized Earth Sciences. *Canadian Journal of Earth Science*, 51, v-viii.

Richards, J.P. 2014. Making faults run backwards: the Wilson Cycle and ore deposits. *Canadian Journal of Earth Science*, 51, 266-271.

West, G.F., Farquhar, R.M., Garland, G.D., Halls, H.C., Morley, L.W. & Russell, R.D. 2014. John Tuzo Wilson: a man who moved mountains. *Canadian Journal of Earth Science*, 51, xvii-xxxi.

Janet Watson

Portrait of Janet Watson, courtesy of Archives Imperial College London.

The Geological Society, housed in Piccadilly, London is doubtlessly one of the world's most prestigious geoscience institutions. Visitors from all around the world attend its conferences and meetings held in the lecture theatre named after one of the greatest geologists of the 20th Century – Janet Watson. This honour is a fitting tribute to one of the most distinguished and well-known personalities in geoscience, famous for her gift of clear and persuasive communication.

Watson was born in 1923 into a geological family. Her father was Professor D.M.S. Watson of University College London, an international authority on vertebrate paleontology. She gained a degree in General Science from Reading University in 1943 going onto Imperial College where she graduated with first class honours in geology in 1947. Imperial was to be her base for the rest of her career collaborating first with the formidable H.H. Read and later with her research student colleague and subsequently husband, John Sutton.

Read was building a research team to study the Precambrian metamorphic rocks of North-West Highlands of Scotland. He was so impressed by Watson's undergraduate performance (he reputably said she should have been given 120% in her finals exams to do her justice) that he invited her to join as a postgraduate student. Her work started on the migmatites of Sutherland (leading to her publishing her first of many scientific papers in 1948), but moved on to the Lewisian of the Scourie area in the remote North-West Highlands. By 1949 she had obtained her PhD and married John Sutton; a honeymoon in the Channel Islands providing an opportunity to start preparations for a one-off paper on the geology of the Isle of Sark! She then began a long career as Research Assistant to Read until 1974 when she became Professor of Geology.

Janet Watson with her husband and scientific collaborator John Sutton in the field, in NW Scotland, early in their research careers. Courtesy of Archives Imperial College London.

Scourie dykes intruding into Lewisian gneiss (lighter rocks) near Loch Laxford in the North-West Highlands of Scotland.

The title Research Assistant underplays what was a major role in a productive scientific relationship with Read. Together they published two classic textbooks – *Introduction to Geology* in 1962 and *Beginning Geology* in 1966. These went through several editions, updated for the revolutions in geoscience taking place in the 1960s and '70s. At the same time she produced numerous papers on Scottish Highland geology, many with John Sutton – the pair being the undoubted dominant figures in this field of research.

The critical contribution from this work was unravelling of the stratigraphy the Lewisian Complex. They recognised two periods of Precambrian Orogeny — the Scourian and Laxfordian — using basic dykes as chronological markers, exemplified in their classic paper in *Geological Magazine* in 1951. Previously the tectonostratigraphy of these basement rocks had seemed impossible to determine, but Watson and Sutton showed that with careful field observation chronostratigraphic subdivision was possible. Subsequent radiometric dating showed that the two orogenies were separated by around a billion years.

These early research activities were followed by studies on various other metamorphic complexes, notably the Moine and Dalradian of Scotland. Watson's innovation in this research was not just the application of careful field observations, but the synthesising of information from multiple disciplines, including sedimentology and structural geology, into a sum greater than its parts. An example would be the recognition of sedimentary structures in the Upper Dalradian succession of Banffshire that indicated turbidite deposition in deep water. Noting the contrast with shelfal facies in the underlying Lower and Middle Dalradian, Watson and Sutton ascribed this change to the onset of Caledonian tectonism and the emergence of a landmass to the northwest, shedding Lewisian-derived material into a deepening Dalradian basin.

Their fieldwork moved to the Outer Hebrides in the 1960s — a return to focus on the Lewisian and its tectonostratigraphic evolution. Watson was supported by a team of research students spread across the islands. She did not drive but covered many miles in these remote locations by bicycle and the local bus service! The result was the development of the definitive set of geological maps of the islands, first published in 1984. The ideas on the Lewisian geology of the type area were exported to Greenland, were she conducted field work in 1975, contributing to her development of a picture of the accretion of the North Atlantic craton.

During her career she became progressively interested in applied geology, notably ore genesis and regional geochemistry. This involved close collaboration with the British Geological Survey, an organisation she had close affinity with given her love of field work. Her last book, published in 1983, demonstrates her growing interests – *Geology and Man – An Introduction to Applied Earth Science*. This research stemmed from her views on the nature of the deep crust where she envisaged within a mobile layer high-grade metamorphism and partial melting leading to the production of granites, migmatites and associated mineralisation.

In 1982 Watson was appointed President of the Geological Society of London, the first woman to hold this role. Previously, in 1973, she had been awarded the prestigious Lyell Medal by the society in recognition of her research (the second woman in the history of the society to be its recipient) and in 1979 elected a Fellow of the Royal Society. In 2016, the Geological Society launched a biennial meeting named in her honour. These are aimed particularly at young professionals, whom she was keen to encourage in their careers. She was a popular speaker at student societies, always being relied upon to give a clear and entertaining presentation, with an emphasis of field geology that is inspiring to students.

Janet Watson passed away in 1985 at the relatively young age of 61 depriving our profession of an influential and inspirational figure. Her legacy can be measured by the number of research students she assisted or supervised in her time at Imperial, many of whom have gone on to their own distinguished careers. Although essentially a private person, she infected people with a genuine enthusiasm for geology, no matter if in the field, the research laboratory or when serving on an international committee.

REFERENCES

This essay has drawn upon information from the following sources:

Bowes, D.R. 1987. Janet Watson – an appreciation and bibliography. In: Park, R.G. & Tarney, J. (eds.) *Evolution of the Lewisian and Comparable Precambrian High Grade Terrains*. The Geological Society, London, Specials Publications, 27, 1-5.

Goodenough, K. M. & Krabbendam, M. 2011. *A Geological Excursion Guide to the North-West Highlands of Scotland*. Edinburgh Geological Society, 215pp.

Fettes, D.J. & Plant, J.A. 1995. Janet Watson. 1 September 1923 – 29 March 1985. *Biographical Memoirs of Fellows of the Royal Society*, 41, 500-514.

Deformation in Lewisian gneiss, Assynt region, North-West Highlands of Scotland.

Permian sediments of the Kapp Starostin Formation on Spitsbergen.

Brian Harland

Many geoscientists would be more than content if they made a notable contribution in one field of geology. The Cambridge-based academic and Arctic explorer Brian Harland made telling contributions in at least three fields, as well as being a remarkable geological "all-rounder," teacher, scientific administrator, and, as his obituary in *The Independent* newspaper noted, "*phenomenally productive investigator of the history of the Earth.*"

Walter Brian Harland (he preferred to use his middle name) was born in Scarborough in 1917. Situated on the Yorkshire coast of England, the local area is noted for its outcrops of Jurassic strata. Whilst still a boy, Harland developed a love for fossil collecting along the Yorkshire coast and at the age of thirteen, he discovered a spectacular partial skeleton of a fossil marine crocodile, *Steneosaurus*. This discovery made the national press and the specimen now resides in The Natural History Museum, London. An auspicious start to the career of a geologist! When sent to boarding school near Malvern near the Welsh Borders, he taught himself how to make a geological map and set about mapping this interesting area of Pre-Cambrian and Paleozoic geology. In 1935, he began his academic geological studies at the University of Cambridge, graduating with first class honours in 1938.

1938 also saw Harland take part in his first research expedition to Spitsbergen, the largest island in the mountainous archipelago of Svalbard in the Norwegian sector of the Arctic Ocean. The geological investigation of these islands would be synonymous with his name — he participated in no less than 43

Brian Harland with expedition supplies at a field base in Spitsbergen in the 1938. Photograph provided by and used with the permission of the Sedgwick Museum of Earth Sciences.

expeditions there, 29 of which he personally led. Many outcrops on Spitsbergen are spectacular and Harland used these not only to geologically map the island, but to provide insight into the tectonostratigraphy of the Arctic and global geological phenomena.

The initial expedition in 1938 was arduous. The six-man team he was part of suffered from severe shortages of food and particularly bad weather. But far from deterring Harland from further work, this only served to instill in him the importance of detailed and meticulous expedition planning. It is not surprising that in the later part of his career he was awarded the Gold Medal of the Royal Geographical Society for his Arctic exploration.

Back at Cambridge, Harland began work on his PhD in 1938, but with outbreak of World War II the following year, his studies were curtailed. As an undergraduate he had become a Quaker, and these religious beliefs led him to work on a farm as a conscientious objector during the early years of the war.

In 1942, he decided to take on a teaching role at Chengdu University in China. He retained an interest in Chinese culture, science and geology for the rest of his life, maintaining links with Chengdu University despite occasional political difficulties. Whilst at Chengdu he developed a close friendship with Joseph Needham, the famous chemist and historian of Chinese science. Both were to become Fellows of Gonville and Caius College in Cambridge.

In 1946, Harland returned to Cambridge to take on a role as a geology demonstrator to undergraduates. This progressively became a lectureship and eventually a readership. He was to be based in Cambridge and Gonville and Caius College for the rest of his career. An enthusiastic teacher, he always emphasized the importance of field geology. His numerous trips leading first-year undergraduates to study the geology of the Scottish island of Arran have passed into Cambridge geology department folklore!

His return to Cambridge also saw a return to exploring the geology of the Arctic, with the initiation of the Cambridge Spitsbergen Expedition programme. This annual programme of geological expeditions led to over 300 graduates and staff from Cambridge having the opportunity to carry out research on and around Spitsbergen. Many of the major names in geoscience today were the beneficiaries of this opportunity. The culmination of decades of research was the publication of the *Geology of Svalbard* in 1997. A landmark in Arctic geology research and vital resource for anyone wishing to unravel the complex geology of this region.

What then of the wider significance of Harland's work on Spitsbergen and on Arctic geology? Comparisons of the geology of Spitsbergen with Greenland led him to be an early advocate of continental drift (later plate tectonics). Moreover, his observations of Pre-Cambrian glacial deposits on Spitsbergen led him to compare and correlate them with similar deposits in other far-flung parts of the globe. This not only provided further fuel for the notion of mobile continents, but also led to the notion that around 600 million years ago, glaciation had been widespread, even at low latitudes. He was an early adopter of paleomagnetic data to show that "infra-Cambrian" (= Late Precambrian) glacial diamictites in Spitsbergen and Greenland were deposited at tropical latitudes. From these data and the sedimentological

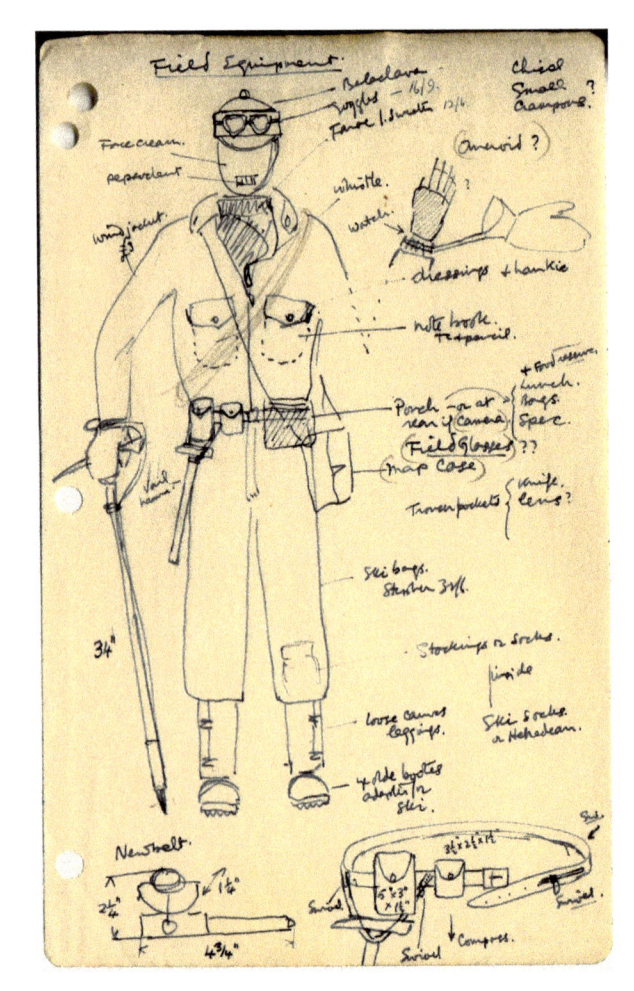

Sketch by Brian Harland of the field gear needed for an early geological expedition to Spitsbergen. This demonstrates the meticulous planning that typified each expedition he led. Image provided by and used with the permission of the Sedgwick Museum of Earth Sciences.

evidence that the glacial sediments interrupt successions of rocks commonly associated with tropical to temperate latitudes, he argued for an ice age that was so extreme that it resulted in the deposition of marine glacial rocks in the tropics. This remarkable hypothesis received little support when Harland first presented it, but is now more widely accepted under the moniker 'Snowball Earth.'

Given the need to correlate the geology of Spitsbergen with other parts of the globe, techniques for correlation and indeed collaboration became central to much of Harland's work from the 1960s onward. Arthur Holmes had published regular updates to his famous geological timescale up until the beginnings of the 1960s. Harland felt that as more data were gathered in the fields of geochronology and biostratigraphy it was imperative that an up-to-date geological timescale was widely available to all researchers. He thus coordinated the research for the book for which he most arguably most well-known — *The Geological Timescale*. It was first published in 1964 with revised and expanded versions in 1971, 1982 and 1990. Not surprisingly, he soon became known as "Timescale Harland" by the geological community at large.

Harland's great skill in creating *The Geological Timescale* book series was his ability to coordinate and synthesize the work of a group of colleagues. He believed deeply in international collaboration and was at the forefront of the founding of the International Geological Correlation Programme, the now-UNESCO funded series of research projects that seek to unravel Earth's history. As Honorary Secretary of the Geological Society of London from 1963 to 1970, he led plans for the Society to become a center for collaborative research. He initiated a series of multi-contribution books, which led to the Society's flagship series of Special Publications that to date includes almost 500 volumes on a diverse range of geological subjects.

As a great 'all-rounder,' Harland's interest in tectonics matched his interest in sedimentology and stratigraphy. He coined the term 'transpression' to describe mountain formation by oblique convergence and introduced the name 'Iapetus' for the Early Paleozoic ocean that ran through parts of modern-day Europe and North America. Previously known by the misnomer Proto-Atlantic, Harland choose the name after a conversation with a classics colleague at Gonville and Caius who pointed out that Iapetus was the mythological father of Atlas, from whom the name Atlantic is derived.

Precambrian diamictite.

Harland's appetite for scientific work can only be marveled at — it is said that he typically worked 14 hours a day. In addition to the activities mentioned, he edited the prestigious journal *Geological Magazine* for 30 years. He edited the first edition of *The Fossil Record*, a compendium of the stratigraphic ranges of all fossil groups.

There is no doubt that the fields of Arctic geology, geochronology, biostratigraphy, tectonics and glacial geology would be much poorer without his contributions and boundless energy to coordinate research activities in these fields.

Brian Harland passed away in 2003. His legacy continues in the form of CASP (formerly the Cambridge Arctic Shelf Programme), an industry-funded charitable research organisation that carries out field work in remote locations.

ACKNOWLEDGEMENTS

Sandra Freshney, Archivist at the Sedgwick Museum of Earth Sciences in Cambridge is thanked for providing some of the images used in this article.

REFERENCES

This essay has drawn on information from the following sources along with the authors recollections of his conversations with Brian Harland and with colleagues who worked alongside him at Cambridge:

Frankel, H.R. 2012. *The Continental Drift Controversy. Volume III: Introduction of Seafloor Spreading.* Cambridge University Press. 476pp.

Friend, P.F. 2004. Walter Brian Harland, 1917-2003. *Proceedings of the Geologists Association*, 115, 183-186.

http://www.independent.co.uk/news/obituaries/w-b-harland-37459.html

http://www.sedgwickmuseum.org/index.php?page=brian-harland-100

Dan McKenzie

Plate tectonics is the defining geoscience paradigm of the second half of the 20th century. 2017 was designated its 50th anniversary by The Geological Society of London because, in 1967, a paper was published that is widely regarded as marking the completion of the initial development of the plate tectonic thesis. This paper, *The North Pacific: An Example of Tectonics on a Sphere,* by Dan McKenzie and Bob Parker, provided the model to describe the translations and rotations on a sphere that, thereby, define plate motions. McKenzie has brought much more to geoscience than just this brilliant paper. He is one of the pre-eminent geophysicists of the late 20th to early 21st century and has made a telling contribution to our understanding of processes operating both within the mantle and the crust.

McKenzie was born in Cheltenham in 1942. His father was a doctor on Harley Street in London, and his mother was noted for her work in garden design and as an author. He attended Westminster School, and after winning a state scholarship in pure and applied mathematics, McKenzie entered King's College, Cambridge in 1960 to study the natural sciences. Mathematics was not considered as being one of the three core scientific subjects in the Natural Science Tripos, so alongside physics and chemistry, McKenzie chose geology after chancing upon Charles Lyell's *'Principles of Geology'* and Archibald Geikie's *'Ancient Volcanoes of Great Britain'* in the school library. Unfortunately, geology was to prove an initial disappointment for McKenzie. He found the lectures dull (especially those that focused on palaeontological classification and zonation), so he dropped the subject at the end of his first year.

Following his graduation, he stayed on at Cambridge as a graduate student working with Edward ("Teddy") Bullard, the pioneering marine geophysicist. McKenzie became interested in how the interior of the earth convects, something completely speculative at that time. He taught himself fluid mechanics and then went to the Scripps Institution of Oceanography at the University of California, San Diego as a visiting scholar. This visit ended hastily after eight months. He had mistakenly travelled on an immigration visa, making him liable for the draft to fight in the Vietnam War. So McKenzie returned to Cambridge, submitting his PhD thesis in 1966, but not before he attended a scientific meeting in New York that was to provide an impetus for his research into plate tectonics.

After he heard lectures by Fred Vine on sea floor spreading and magnetic anomalies, McKenzie applied his knowledge of thermodynamics to the problem of how plates move and came up with a model that demonstrated a far more dynamic Earth than anyone had previously thought. He suggested there are two layers in the mantle, each of which is in motion, controlling the movement and behaviour of the tectonic plates above. He published a key paper, *The Viscosity of the Lower Mantle*, in 1966.

Dan McKenzie in the Geological Society Library in 2017. Photograph reproduced with permission of The Geological Society.

Prior to 1967, plate tectonics had a long gestation period, beginning with Alfred Wegener's continental drift (if not the ideas of others before him), and required a whole series of observations during the late 1950s and early '60s, including data on the bathymetry of the deep ocean floors and the nature of the oceanic crust. More generally, the development of marine geology gave evidence for the association of seafloor spreading with mid-oceanic ridges and magnetic field reversals, as published between 1959 and 1963 by Heezen & Tharp, Dietz, Hess,

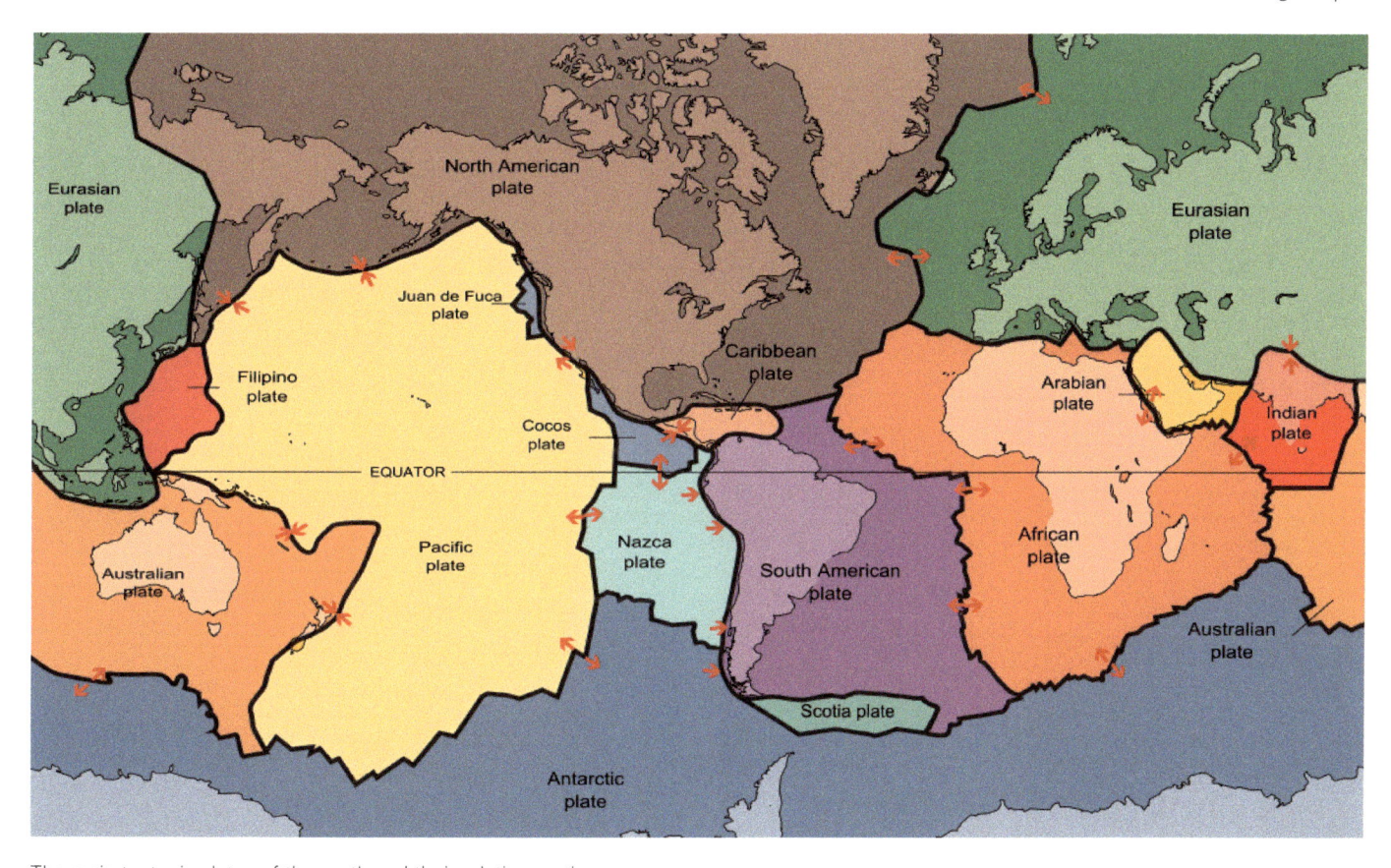

The main tectonic plates of the earth and their relative motions.

Mason, Vine & Matthews, Morley, and others.

Simultaneous advances in seismic imaging techniques along the trenches bounding many continental margins, together with many other geophysical (e.g., gravimetric) and geological observations, showed how the oceanic crust could be subducted, providing the mechanism to balance the extension of the ocean basins with shortening along its margins. The global seismic network and the ability to locate earthquakes and define focal mechanisms was also critical for understanding the geometry of plates, mid-ocean ridges, and the nature of transform faults, all in turn of paramount importance to the development of the plate tectonics paradigm.

By 1967, all that was required was a model describing translations and rotations on a sphere to define plate motions. This was independently and almost simultaneously proposed by McKenzie and Parker, and by the American, Jason Morgan. In *The North Pacific: An Example of Tectonics on a Sphere,* McKenzie employed Euler's Fixed Point Theorem, in conjunction with magnetic anomalies and earthquakes foci to determine a precise mathematical theory for plate tectonics. Tuzo Wilson had suggested in 1965 that the surface of the Earth could be divided into rigid aseismic regions (i.e., plates). In the previous year, Teddy Bullard had used Euler's theorem to describe rigid movements on a sphere when he made continental reconstructions. McKenzie combined the two concepts, which became the modern theory of plate tectonics.

This work was published some months after (unknown at the time to McKenzie) similar ideas had been presented by Jason Morgan of Princeton at an American Geophysical Union conference. McKenzie and Parker's published paper appeared in the journal Nature at the end of 1967. Morgan's published paper appeared early in 1968. The papers are in fact complementary. McKenzie and Parker focused on the earthquake data, while Morgan focused on the magnetic anomalies.

In the late 1960s, McKenzie published a succession of papers on topics key to refining the plate tectonics paradigm. These included convection modelling and vector analysis of triple junctions. In 1968, he began work with John Sclater on surveying the Indian Ocean with the intent of determining its entire geological history. The resultant landmark publication in 1971 eventually resulted in both authors receiving Fellowships at the Royal Society.

At this point, McKenzie decided to move away from plate tectonics, marine geology, and geophysics, choosing instead to focus on the behaviour of fluids below the plates. He studied cellular convection and motions in the mantle, using his trademark approach of identifying the fundamental physical considerations and testing them in the simplest possible numerical models. His work demonstrated that mantle convection would produce a measurable gravity signal, thus, testable by observation.

The early 1970s saw McKenzie investigating tectonic processes within plates, aseismic regions that required a model to explain how extensional basins form. Since these are basins where oil and gas are often discovered, this work (the "McKenzie Model") has been of profound importance to the oil industry. Leading Equinor geologist Tony Doré described the resultant 1978 paper (modestly titled *Some Remarks on the Development of Sedimentary Basins*) as the most important for hydrocarbon exploration in recent decades. Having collected data in the Aegean region, McKenzie understood stretching was taking place not only in the crust, but in the mantle, too. Fault-controlled subsidence is followed by thermal subsidence. Stretching factors can be calculated along with associated geothermal history. This is of fundamental importance in determining the maturity of source rocks within a basin.

This work was a natural progression from another important (and also modestly titled!) 1967 paper, *Some Remarks on Heat Flow and Gravity Anomalies,* which presented an explanation for the observations of gravity, bathymetry, and heat flow at oceanic ridges. The conceptual leap was to apply the same physical processes of cooling of the oceanic lithosphere with time to the post-rift development of sedimentary basins.

McKenzie continues to work at the Bullard Laboratories in Cambridge, where he is Emeritus Professor of Earth Sciences. Most recently, his research has provided new insights into the tectonic evolution of Mars and Venus. He has always been at the forefront of new technologies and sources of data. Just as earthquake seismology provided a key to understanding plate motions fifty years ago, he now sees continuing developments in seismology, such as full waveform inversion, as key to revealing more of the structure of the earth, from crust to core.

In a recent interview, he remarked, "*I wondered for a long time whether I was going to be one of those scientists who has one really good idea when they're in their twenties and that is it.*" Far from one *brilliant* idea, McKenzie has had many, encompassing plate tectonics, mantle dynamics, sedimentary basin formation, melt generation, and planetary geology. Needless to say, he has received many honours, including receiving the Copley Medal of the Royal Society and the Crafoord Prize of the Royal Swedish Academy of Sciences.

REFERENCES

This essay has drawn upon information from the following sources:

https://www.mckenziearchive.org/

https://www.newscientist.com/article/mg23631530-600-dan-mackenzie-the-man-who-made-earth-move/

Allen, P.A. & Allen, J.R. 2005. *Basin Analysis*. Blackwell, 549pp.

Bowler, S. 2017. Fifty years on. *Geoscientist*, October 2017, 10-15.

Frankel, H.R. 2012. *The Continental Drift Controversy. Volume 4 Evolution into Plate Tectonics*. Cambridge University Press, 476pp.

McKenzie, D. 2003. Plate tectonics: a surprising way to start a scientific career. In: Oreskes, N. (ed.) *Plate Tectonics: An Insiders History of the Modern Theory of the Earth*. Westview Press, 169-190.

Peter Vail – Photograph courtesy of Northwestern University.

Peter Vail

There is little doubt that the concepts of sequence stratigraphy have revolutionized both sedimentary geology and petroleum geology over the last forty years. Sequence stratigraphy examines the stratigraphic geometries and associated patterns of sedimentary facies that are generated by relative sea-level change. In doing so, it is a valuable tool for predicting the occurrence of, for example, reservoir and source rock facies and understanding the architecture of reservoirs. Of equal importance, it provides a catalyst for the integration of seismic and well and outcrop data in addition to detailed sedimentological, biostratigraphical and geochemical studies. Put simply, many of the deep-water plays being explored for today are lowstand fans, predicted from sequence stratigraphic principles and identified on high-resolution seismic data.

The sequence stratigraphic methodology first came to prominence with the publication of seminal papers by Peter Vail and his colleagues from Exxon in 1977. Since then, sequence stratigraphic studies have become commonplace and the science has developed its own particular jargon to account for the countless ways in which sediments respond to sea-level change. But Vail and his colleagues did not just bring sequence stratigraphy to the petroleum geologists' tool box; they reactivated an old idea that some sea-level changes are synchronous and global in nature. Such changes are termed eustatic, as first introduced by the great Austrian geologist Eduard Suess in 1888. This has tremendous predictive value, although the recognition of eustasy in the rock record has both ardent supporters and critical skeptics.

Peter Vail was born in 1930 in New York. After attending Dartmouth College, New Hampshire he earned his Master of Science and PhD degrees at Northwestern University, Illinois. Following this in 1956 he joined the Carter Oil Company in Tulsa, Oklahoma. This company was an affiliate of Exxon, with whom he would spend much of his career, until retirement in 1986 when he joined the faculty at Rice University in Houston.

With an academic background in stratigraphic mapping, and influenced by two of the great professors at Northwestern University, Larry Sloss and Walter Krumbein, his initial tasks at Carter Oil were subsurface mapping and correlation projects for exploration in the Paradox Basin, Illinois Basin and Venezuela. This led him to consider what were the right units to map and the importance of time in stratigraphy. In his view, log correlations needed to be placed in time context, not simply be lithological. Correlation of facies changes and maps of geological time slices could be more valuable than maps of lithostratigraphy. This was the beginning of a strong association with biostratigraphers who could provide the important age calibration and constraint.

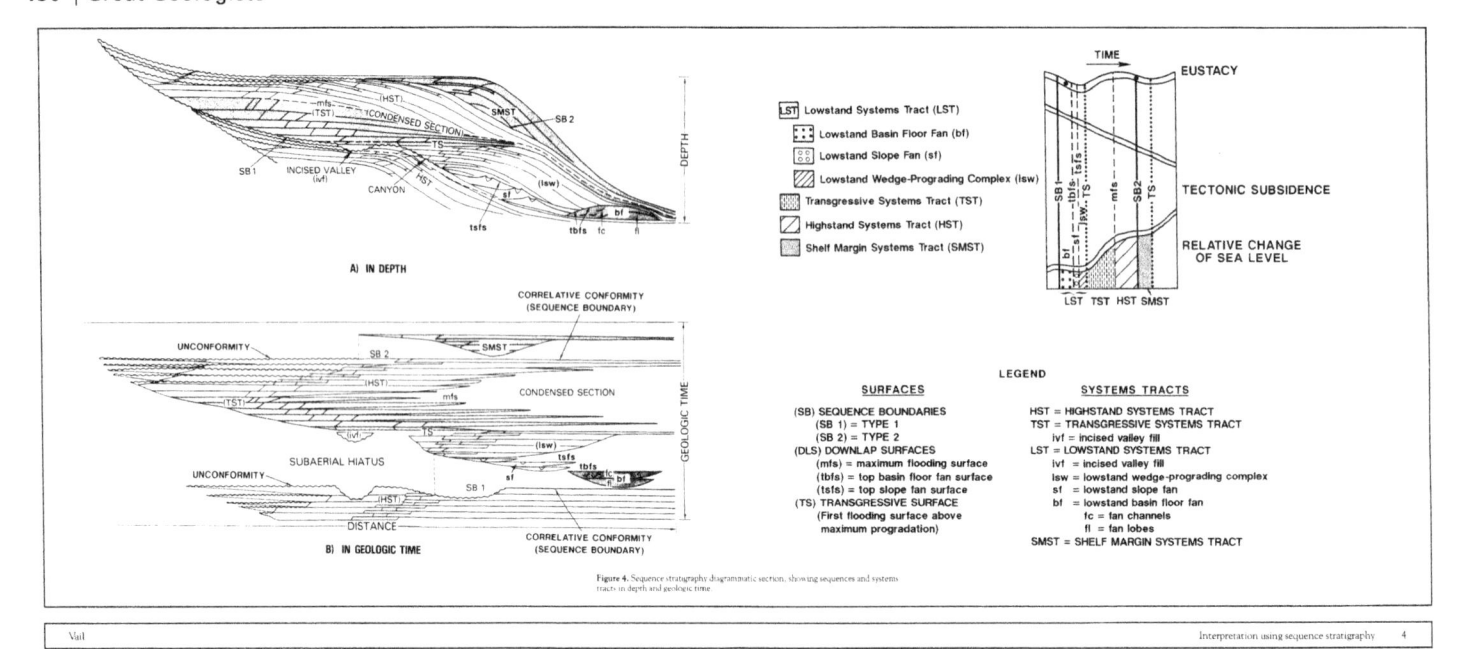

A version of the classic "slug diagram" (from AAPG Studies in Geology 27, 1987) illustrating Vail's concepts of stratal geometries as seen in seismic data and how they can be interpreted in terms of chronostratigraphy. AAPG ©1987, reprinted by permission of the AAPG whose permission is required for further use.

By the early 1960s, seismic interpretation had become an important component of Vail's pursuit of time-based correlation and mapping. Against the advice of colleagues he joined a geophysics team. A project integrating wells, seismic and biostratigraphy from Guinea Bissau led to an important revolutionary revelation – many seismic reflections did not follow time transgressive formational boundaries but instead followed bedding patterns or the real physical surfaces in the rock. In other words they are time lines. Prior to this seismic reflectors had been thought to solely represent impedance contrasts at lithological boundaries. Vail would subsequently write *"primary seismic reflectors are generated by stratal surfaces which are chronostratigraphic, rather than by boundaries of arbitrary defined lithostratigraphic units"*. This new understanding (although still debated today) elevated seismic data to being a new tool in the methodology to make regional chronostratigraphic frameworks for mapping and understanding palaeogeography, especially when coupled with well logs and biostratigraphy. Moreover, the depositional geometries displayed within seismic data gave insight into the pattern of relative sea-level change through recognition of onlap, downlap, etc.

Vail examined seismic data from various parts of the world and was struck by the apparently synchronous nature of many of the relative sea-level changes seen in the sections he studied, calibrated by biostratigraphic data from wells. In 1959 he had drawn his first Phanerozoic sea-level cycle chart based solely on well log correlations. In 1963 he presented this at a company forum augmented by seismic data and in 1966 publically presented his ideas on eustasy at the AAPG annual convention in St. Louis. At this time he was actively using seismic/sequence stratigraphy to guide exploration for Exxon in the newly licensed North Sea.

By his own account, not all his peers in Exxon were enamored with these geoscience breakthroughs. He recounts that at one internal research conference a senior member of staff goaded the audience into laughter after a presentation by Vail with suggestions that he was implying that seismic reflectors must bounce back off fossils. But Vail was not disheartened and persisted with his research eventually being given the opportunity to lead a small research team of geologists, geophysicists and computer scientists working on seismic stratigraphy. As much effort was put into processing seismic

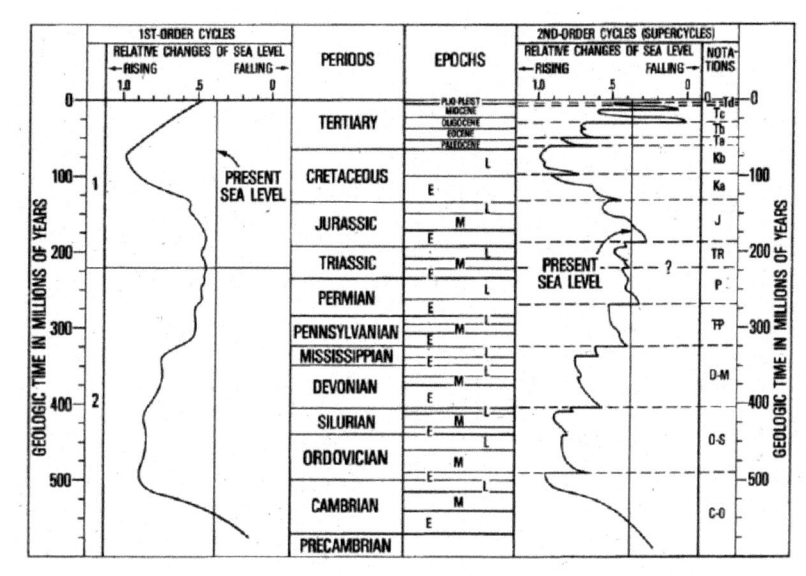

Peter Vail's overview of Phanerozoic eustasy from AAPG Memoir 26 in 1977. AAPG ©1977, reprinted by permission of the AAPG whose permission is required for further use.

data to make it more useful as was interpreting it. The results started to express themselves in exploration success stories – the identification of the Balder lowstand fan in the North Sea for example.

By the 1970s Vail was increasingly involved in outreach activity and began to deliver a Seismic Stratigraphy School for AAPG with the lecture notes eventually released as the seminal AAPG Memoir 26 "*Seismic Stratigraphy – Applications to Hydrocarbon Exploration*" in 1977. The integrative and predictive power of sequence stratigraphy and eustasy now began to be embraced by the exploration industry at large. Academia also began to take notice of sequence stratigraphic concepts and geologists working on outcrops began in earnest to recognize the sequences and eustatic changes mentioned in AAPG Memoir 26 and subsequent publications. The 'sequence stratigraphy industry' of the late 20th century that occupied a significant proportion of geoscience academia striving to relate particular outcrops to the eustatic models of Vail and his colleagues has been criticized by some. There were doubtlessly cases of force-fitting and circular reasoning from some academics. But Vail always highlighted the need to judge the evidence in the rocks on its own merits. During a conference field trip in Europe during the 1980s Vail was repeatedly asked at every outcrop "is this your late Valanginian sea-level fall", "where exactly is the sequence boundary" or similar. To his credit he replied look at the rocks and judge the evidence for yourself. In the later phase of his career he has promoted this wise advice in his teaching at Rice University and with his numerous academic collaborations to advance sequence stratigraphic and eustatic models.

REFERENCES

Vail, P.R. 1992. The evolution of seismic stratigraphy and the global sea-level curve. In: Dott, R.H. Jr. (ed.) *Eustasy: The Historical Ups and Downs of a Major Geological Concept.* Geological Society of America Memoir, 180, 83-92.

https://archives.aapg.org/explorer/2003/05may/slc_vail.cfm

https://www.youtube.com/watch?v=Wpz7tFjxTHg

https://www.youtube.com/results?search_query=%22peter+vail%22+%22mini+geology%22+interview

The Tibetan Plateau, uplift of which has been a major factor in controlling global climates since the Miocene.

Maureen Raymo

The Wollaston Medal is the highest annual accolade of the Geological Society of London. It was established by William Hyde Wollaston, a noted British chemist, to promote *'researches concerning the mineral structure of the Earth' ...'or of the science of Geology in general'*. This medal is given to geoscientists who have made a significant impact through the publication of a substantial body of excellent research. It was first awarded to William Smith in 1831 and the subsequent roll of honour is a listing of pre-eminent geologists. Remarkably, it was not until 2014, 183 years after the award to Smith that the medal was first awarded to a woman — Maureen Raymo.

Currently based at the Lamont-Doherty Earth Observatory of Columbia University in New York, Raymo has been one of the most influential paleoclimatologists of the last 30 years. Her work encompasses three main strands. Firstly, she co-authored the 'uplift-weathering hypothesis,' which linked the creation of the Himalayas and the Tibetan Plateau with the onset of global cooling and polar ice sheet expansion, demonstrating a causal link between global climate and tectonic processes. Secondly, she pioneered the synthesis of oxygen isotope data from benthic records to establish a standard proxy for sea-water temperature and ice volume variations over the last five million years. Thirdly, she created and led the Pliomax project which led to a better understanding of sea-level and climate change during the Pliocene. The mid-Pliocene is a good analogue for a future Earth subjected to substantial global warming and elevated atmospheric CO_2. Her 2008 review article in *Nature* with Peter Huybers titled '*Unlocking the mysteries of the ice ages*' is a testament to the mark she has made in understanding glacial-interglacial climate changes and her ability to tackle some of the most long-standing problems in this field.

Born in Los Angeles, Raymo was raised near Boston, in Easton, Massachusetts, by a father who taught college physics and wrote popular science books, and a mother who taught children with learning disabilities. She first became interested in science as a seven-year-old watching the popular television show The Undersea World of Jacques Cousteau. She has said, *"When I was a kid I wanted to be an oceanographer like Jacques Cousteau. The only time I drifted away from that goal was in high school, when I got into geology."* But as a college student at Brown University, she discovered a field that combined oceanography and geology: paleoclimatology.

Her interest in palaeoclimatology led her to accept a Ph.D. at Columbia University researching the initiation of the great ice sheets in the northern hemisphere during the Cenozoic. This was followed by spells at a number of leading academic institutions including the University of Melbourne in Australia, the University of California at Berkeley, Boston University and MIT, followed by return to Columbia in 2011.

It was as graduate student of Bill Ruddiman at Columbia that she rose to prominence when they and co-author Philip Froelich proposed the 'uplift-weathering hypothesis' in 1988. This established a link between the Late Eocene uplift of the Himalayas and Tibetan plateau and the global cooling trend seen in the Cenozoic. Since the Late Miocene, uplift rates have been particularly pronounced, and polar ice sheets have episodically expanded

Maureen Raymo in her office at Columbia University. Image used with her permission.

The Wollaston Medal of the Geological Society of London. The medal, awarded to outstanding geologists, is cast in Palladium, an element discovered by Wollaston.

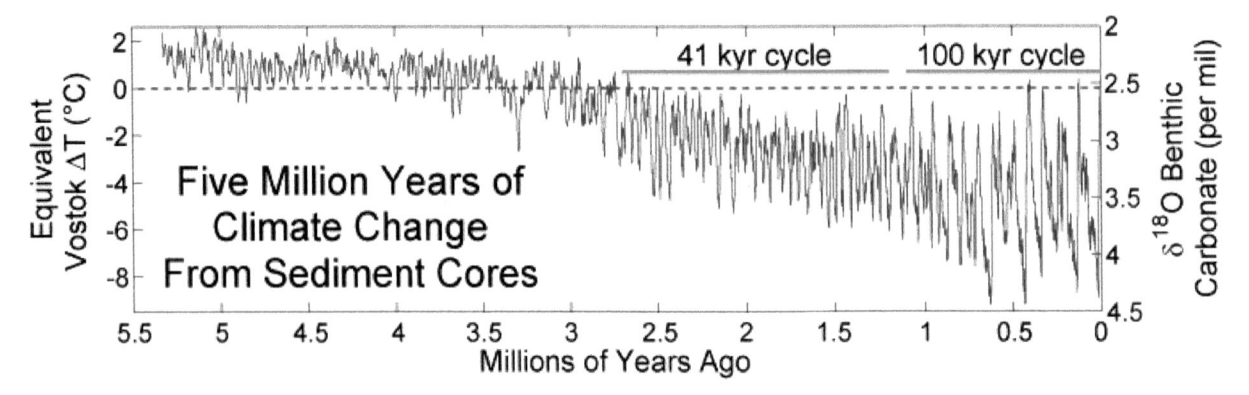

Five Million Years of Climate Change From Sediment Cores

A version of the LR04 Stack created by Maureen Raymo and Lorraine Lisiecki summarizing benthic stable oxygen isotope records for the last 5.3 Ma and hence paleotemperature and ice volume. Note the variations in periodicity.

extensively. This link, they hypothesized, related to the intensification of the Indian monsoon, which acted to increase weathering, drawing down atmospheric CO_2. This is arguably the most important idea in Cenozoic climate change to emerge in the last 30 years and continues to be vigorously debated today.

A major criticism of the hypothesis was that chemical weathering rates on land could not increase in the absence of enhanced metamorphic delivery of CO_2 to the atmosphere (for which there is no evidence), otherwise CO_2 would be completely stripped from the atmosphere within a few hundred thousand years and the Earth would become a frozen "snowball" planet. Raymo and Ruddiman agreed that a negative feedback was needed to stabilize atmospheric CO_2 levels and argued that this feedback may have operated through the organic carbon subcycle.

Raymo was central to many of the most significant developments in the astronomical/orbital interpretation of deep-sea oxygen isotope records derived from foraminifera tests, and what they tell us about Cenozoic climate and sea-level change. This work culminated in her seminal 2005 publication, with her graduate student Lorraine Lisiecki, of the iconic 'LR04 Stack' of benthic oxygen isotope records covering the last 5.3 million years, which measures global ice volume and deep ocean temperature.

The LR04 Stack provided the paleoclimate community with two widely applicable stratigraphic tools: a common timescale; and a correlation target for the vast number of paleoceanographic records that have since been collected worldwide. Not surprisingly, this work has been cited over 4000 times.

From the LR04 Stack one can see that the periodicity of glaciations was more frequent between 3 Ma and 800,000 years ago than more recently, occurring every 41,000 years instead of every 100,000. The reason for this shift has long remained a mystery though Raymo has proposed an explanation, the "Anti-phase Hypothesis". From her work, it appears that the Antarctic ice sheet was far more dynamic in the past than generally accepted.

With atmospheric CO_2 concentrations c. 400 ppm, Raymo realized that the mid-Pliocene is one of the most important analogues for an anthropogenically-warmed future. She led the international Pliomax project to investigate and refine our understanding of global sea levels during this period. This has led to seminal collaboration with geophysicists, notably using glacial-isostatic adjustment models to establish Pliocene sea level, and comparing these with the observational database. In addition to solving how to correct ancient shorelines for subsequent movement due to glacial isostatic adjustment, a major contribution of the PLIOMAX project has been demonstrating the pervasive nature of vertical deformation on so-called tectonically passive margins. These vertical deflections are caused by dynamic topography driven by vertical motions in the mantle.

Raymo and her colleagues have visited Australia, Italy, South Africa, the eastern United States and, Patagonia to measure past sea-level high stands. After correcting for shoreline movement from tectonic activity and the loading and unloading of ancient ice sheets, she and her team are amassing a continent-by-continent map of where the seas stood. Ultimately, they hope to understand to what extent melting from Greenland and Antarctica contributed to sea-level rise when global average temperatures were 2–3°C warmer than today.

Raymo's many publications range from the detailed science in high-impact peer-reviewed journals to more general pieces aimed at a wider audience. Communication of science has always been high on her agenda and she has been frequently interviewed by the media, both to explain how the study of past climates can inform an understanding of future climate change and as a leading female scientist. *Written in Stone — a Geological History of the Northeastern United States* co-authored with her father is a well-received book aimed at the interested layperson.

The Wollaston Medal is just one of many accolades awarded to Raymo. In 2014 she received the Milutin Milankovic Medal at the European Geosciences Union's annual meeting in April for marshalling evidence from geochemistry, geology and geophysics to solve paleoclimatology's big problems. "*In an age of progressive research specialization, Raymo remains a model of interdisciplinary research*" reads the award citation. In 2016 she was elected a member of the National Academy of Sciences. Her citation for the Wollaston Model notes that "*you are a profoundly important figure in marine geology, paleoceanography, climate and Earth System science. Your ideas and creativity have set the agenda for 30 years.*" A fitting tribute to an extraordinary geologist.

REFERENCES

This essay has drawn upon information from the following sources:

https://www.geolsoc.org.uk/About/History/Awards-Citations-Replies-2001-Onwards/2014-Awards-Citations-replies

http://www.bu.edu/bridge/archive/2003/09-26/raymo.html

https://www.ldeo.columbia.edu/news-events/climate- scientist-first-woman-win-geologys-storied-wollaston-medal

https://www.egu.eu/awards-medals/milutin-milankovic/2014/maureen-e-raymo/

http://moraymo.us/projects

GREAT GEOLOGISTS: PICTURE CREDITS

Unless specified, images used are in the public domain and/or reproduced under a Creative Commons license. In the case of copyrighted images they are used with the permission of the copyright owner as noted below. Any images not falling into these categories are copyrighted to Halliburton or the author as noted below.

Page 12: https://commons.wikimedia.org/wiki/File:Portrait_of_Nicolas_Stenonus.jpg

Page 13: ©Halliburton

Page 14: https://en.wikipedia.org/wiki/Abraham_Gottlob_Werner#/media/File:Abraham_Gottlob_Werner.jpg

Page 16: https://commons.wikimedia.org/wiki/File:Scheibenberg,_Sachsen_-_Stadtansicht_(Zeno_Ansichtskarten).jpg

Page 17: https://commons.wikimedia.org/wiki/File:Freiberg_Brennhausgasse_14_Werner-Bau.jpg

Page 18: https://commons.wikimedia.org/wiki/File:Sir_Henry_Raeburn_-_James_Hutton,_1726_-_1797._Geologist_-_Google_Art_Project.jpg

Page 19 top: https://commons.wikimedia.org/wiki/File:Siccar_point_red_capstone_from_above.jpg

Page 19 bottom: https://commons.wikimedia.org/wiki/james_hutton#/media/file:james_hutton_field.jpg

Page 20: ©Prof. Iain Stewart, University of Plymouth and used with his permission

Page 21: https://en.wikipedia.org/wiki/William_Smith_(geologist)#/media/File:Geological_map_Britain_William_Smith_1815.jpg

Page 22: https://en.wikipedia.org/wiki/William_Smith_(geologist)#/media/File:William_Smith_(geologist).jpg

Page 23 top: https://en.wikipedia.org/wiki/William_Smith_(geologist)#/media/File:Geological_map_Britain_William_Smith_1815.jpg

Page 23 bottom: Smith, W., 1815. A Memoir to the Map and Delineation of the Strata of England and Wales with Part of Scotland. John Cary. http://images.shoutwiki.com/chester/9/9a/Smithsection.jpg

Page 24: https://en.wikipedia.org/wiki/Georges_Cuvier#/media/File:Georges_Cuvier.png

Page 25: Collini, C A. 1784. Sur quelques Zoolithes du Cabinet d'Histoire naturelle de S. A. S. E. Palatine & de Bavière, à Mannheim. Acta Theodoro-Palatinae Mannheim 5 Pars Physica, 58–103. http://www.wikiwand.com/en/Pterodactylus

Page 26: Cuvier, G. & Brongniart, A. 1811. Essai sur la géographie minéralogique des environs de Paris: avec une carte géognostique, et des coupes de terrain. Baudouin

Page 28: https://commons.wikimedia.org/wiki/User:John_Cummings

Page 29 left: ©Look and Learn and used with their permission

Page 29 center: https://en.wikipedia.org/wiki/Mary_Anning#/media/File:Mary_Anning_Plesiosaurus.jpg

Page 29 right: Buckland, W. 1836. Geology and Mineralogy Considered with Reference to Natural Theology, William Pickering, London. https://commons.wikimedia.org/wiki/

File:Anning_plesiosaur.png

Page 30-31: Buckland, W., 1836. Geology and Mineralogy Considered with Reference to Natural Theology, William Pickering, London.

Page 31: Buckland, W., 1824. Notice on the Megalosaurus or great Fossil Lizard of Stonesfield. Transactions of the Geological Society of London, 2(2), 390-396. https://commons.wikimedia.org/wiki/File:Buckland,_Megalosaurus_jaw.jpg

Page 32: http://wellcomeimages.org/indexplus/obf_images/38/ee/f7f33f27ea79570dff7399fe7d17.jpg. by/4.0/deed.en

Page 33: https://en.wikipedia.org/wiki/Trinity_College,_Cambridge#/media/File:Cmglee_Cambridge_Trinity_College_Great_Court.jpg

Page 34: https://en.wikipedia.org/wiki/Adam_Sedgwick#/media/File:Adam_Sedgwick.jpg

Page 35: The author

Page 36: https://commons.wikimedia.org/wiki/File:Sir_Roderick_Impey_Murchison,_1st_Bt_by_Stephen_Pearce.jpg

Page 37: Murchison, R.I., 1835. On the Silurian system of rocks. The London, Edinburgh, and Dublin Philosophical Magazine and Journal of Science, 7(37), 46-52. https://commons.wikimedia.org/wiki/File:Cross-Section_of_Silurian_System.png

Page 38 left: https://www.geolsoc.org.uk/~/media/CD0ED1B3AB914058945D591CAC0F0B55.ashx

Page 38 right: http://www.geowestmidlands.org.uk/wiki/index.php5?title=File:Dudleybug.jpg

Page 39: https://en.wikipedia.org/wiki/File:Lyell_1840.jpg

Page 40: Lyell, C. 1838. Elements of Geology, J. Murray. https://upload.wikimedia.org/wikipedia/commons/1/1a/Lyell_Principles_frontispiece.jpg

Page 41 left: https://commons.wikimedia.org/wiki/File:Ichthyosaurs_attending_a_lecture_on_fossilised_human_remains_Wellcome_V0001518.jpg

Page 41 right: Lyell, C. 1830. Principles of Geology, Being an Attempt to Explain the Former Changes of the Earth's Surface, by Reference to Causes Now in Operation. J. Murray. https://en.wikipedia.org/wiki/Principles_of_Geology#/media/File:Charles_Lyell_-_Pillars_of_Pozzuoli.jpg

Page 42: Courtesy of Dr. Geraint Wyn Hughes.

Page 43: https://en.wikipedia.org/wiki/Alcide_d%27Orbigny#/media/File:Alcide_Dessalines_d%27Orbigny_1802.jpg

Page 44 left: d'Orbigny, A.D.1826. Tableau méthodique de la classe des Céphalopodes. Annales des Sciences Naturelles. Series, 1, 96-314. https://paleonerdish.files.wordpress.com/2014/08/sin-tc3adtulo3.jpg

Page 44 right: d'Orbigny, A.D., 1842. Paléontologie française. Terrains jurassiques. 1. Céphalopodes, Éditions Masson, Paris. http://www.bajocien14.com/article-discohelix-sinistra-99365428.html

Page 47: https://en.wikipedia.org/wiki/File:Charles_Darwin_by_G._Richmond.jpg

Page 48: Darwin, C.R. 1846. Geological Observations on South America: Being the Third Part of the Geology of the Voyage of the Beagle, Under the Command of Captain Fitzroy, RN During the Years 1832 to 1836. Smith, Elder and Company, 65pp. http://darwin-online.org.uk/content/frameset?itemID=F273&viewtype=image&pageseq=283

Page 49: Darwin, C. 1842. On the Structure and Distribution of Coral Reefs: Being the First Part of the Geology of the Voyage of the Beagle Under the Command of Captain Fitzroy, RN During the Years 1832 to 1836. Smith, Elder and Company. http://darwin-online.org.uk/content/frameset?itemID=A622&viewtype=text&pageseq=37

Page 51: https://en.wikipedia.org/wiki/Louis_Agassiz#/media/File:Louis_Agassiz_H6.jpg

Page 52: https://commons.wikimedia.org/wiki/File:Unteraargletscher.jpg

Page 53: Agassiz, L. 1833–1844. Recherches sur les poissons fossiles, 5 vols. Imprimiere de Petipierre, Neuchatel, 1-336. https://darwin.lindahall.org/44_agassiz_a.shtml

Page 54 left: Agassiz, L. 1833–1844. Recherches sur les poissons fossiles, 5 vols. Imprimiere de Petipierre, Neuchatel, 1-336.

Page 54 right: Agassiz, L. 1840. Etudes sur les glaciers. Jent et Gassmann.

Page 55: https://en.wikipedia.org/wiki/James_Dwight_Dana#/media/File:James_Dwight_Dana_by_Daniel_Huntington_1858.jpeg

Page 56: Dana, J.D. 1834. On the conditions of Vesuvius in July, 1834. American Journal of Science and Arts, 1st ser., 27, 281–288. http://etc.usf.edu/clipart/45100/45105/45105_mt_vesuvius.htm

Page 57: ©Halliburton

Page 58: https://inkct.com/2017/06/david-friend-hall/

Page 59: https://de.wikipedia.org/wiki/Charles_Lapworth#/media/File:Charles_lapworth.jpg

Page 60 left: The author

Page 60 right: https://en.wikipedia.org/wiki/Dob%27s_Linn#/media/File:Climacograptus_wilsoni_Graptolite_Fossils_from_Dob%27s_Linn_Scotland.jp

Page 61: ©The Lapworth Museum of Geology and used with their permission

Page 62: ©The Natural History Museum and used with their permission

Page 63: ©The Lapworth Museum of Geology and used with their permission

Page 64: ©The University of Sheffield and used with their permission

Page 65: The author

Page 66: The author

Page 67: https://en.wikipedia.org/wiki/Ripple_marks#/media/File:Climbing_ripples.JPG

Page 68: https://he.wikipedia.org/wiki/%D7%A7%D7%95%D7%91%D7%A5:Folding_Gasteretal.jpg

Page 69 left: https://commons.wikimedia.org/wiki/File:Eduard_Suess_1869.jpg

Page 69 right: https://en.wikipedia.org/wiki/File:Tethys_mosaic_83d40m_Phillopolis_mid4th_century_-p2fx.2.jpg

Page 71: https://commons.wikimedia.org/wiki/File:PSM_V51_D808_Thomas_Chrowder_Chamberlin.jpg

Page 72: Chamberlin, T.C. 1894. Glacial phenomena of North America. In: Geikie, A. (ed.) The Great Ice Age, 3rd Edition, Stanford, London, 724-775.

Page 73: ©Professor James Aber and reused with his permission

Page 74: https://upload.wikimedia.org/wikipedia/commons/f/fa/PSM_V51_D224_Alexander_Karpinsky.png

Page 75 top: https://en.wikipedia.org/wiki/Russian_Academy_of_Sciences#/media/File:Russian_Academy_of_Sciences_SPB.jpg

Page 75 bottom left: https://en.wikipedia.org/wiki/Helicoprion#/media/File:Helicoprion_tooth_whorl.jpg

Page 75 bottom right: https://en.wikipedia.org/wiki/Helicoprion#/media/File:Helicoprion_NT_small.jpg

Page 76: https://en.wikipedia.org/wiki/Ural_Mountains#/media/File:Gorskii_04428u.jpg

Page 77: ©Iain Sarjaent and reused with his permission

Page 78: https://upload.wikimedia.org/wikipedia/commons/b/b9/Peach_and_Horne.jpg

Page 79 left: ©Halliburton

Page 79 right: ©Halliburton

Page 80 left: Courtesy of Prof. Rob Butler, University of Aberdeen

Page 80 right: Peach, B.N., Horne, J., Gunn, W., Clough, C.T., Hinxman, L.W. & Teall, J.J.H. 1907. The Geological Structure of the North-West Highlands of Scotland. Memoirs of the Geological Survey of Great Britain.

Page 81: ©British Geological Survey and reproduced with their permission. http://geologicalnotes.weebly.com/uploads/1/1/5/6/11564094/412959.jpg?555

Page 82: ©Halliburton

Page 83: https://en.wikipedia.org/wiki/Alfred_Wegener#/media/File:Alfred_Wegener_ca.1924-30.jpg

Page 85: https://en.wikipedia.org/wiki/Arthur_Holmes#/media/File:Arthurholmesin1912.jpg

Page 86: Redrawn after Holmes, A. 1947. The construction of a geological time-scale. Transactions Geological Society of Glasgow, 21, 117-152.

Page 87: Redrawn after Holmes, A. 1944. Principles of Physical Geology. Thomas Nelson and Sons, Ltd.

Page 89: https://en.wikipedia.org/wiki/Milutin_Milankovi%C4%87#/media/File:Milutin_Milankovi%C4%87.jpg

Page 90: https://en.wikipedia.org/wiki/Milutin_Milankovi%C4%87#/media/File:Milutin_Milankovi%C4%87_2004_Serbian_stamp.jpg

Page 91: ©Halliburton

Page 92: Petit, J.R., et al. 2001. Vostok Ice Core Data for 420,000 Years, IGBP PAGES/World Data Center for Paleoclimatology Data Contribution Series #2001-076. NOAA/NGDC Paleoclimatology Program, Boulder CO, USA.

Page 93: The author

Page 94: https://en.wikipedia.org/wiki/Amadeus_William_Grabau#/media/File:Amadeus_William_Grabau.jpg

Page 95: ©Halliburton

Page 97: ©The Royal Society and reproduced with their permission

Page 98: https://en.wikipedia.org/wiki/Dactylioceratidae#/media/File:Ammonites_Dactylio_Ceras_Commune_Schleifhausen_Germany_Jura_02.jpg

Page 99: The author

Page 101 left: ©Lamont-Doherty Earth Observatory and the estate of Marie Tharp and reproduced with their permission

Page 101 right: https://en.wikipedia.org/wiki/Marie_Tharp#/media/File:Tharp_%26_Heezen.jpg

Page 102: World Ocean Floor Panorama, Bruce C. Heezen and Marie Tharp, 1977. Copyright by Marie Tharp 1977/2003. Reproduced by permission of Marie Tharp Maps LLC.

Page 103: World Ocean Floor Panorama, Bruce C. Heezen and Marie Tharp, 1977. Copyright by Marie Tharp 1977/2003. Reproduced by permission of Marie Tharp Maps LLC.

Page 104: https://en.wikipedia.org/wiki/Ziad_Rafiq_Beydoun#/media/File:Don_Beydoun_1960.jpg

Page 105: https://scoopempire.com/photos-remind-beautiful-yemen/anthonypappone-wadi-doan-hadramawt/

Page 106: https://en.wikipedia.org/wiki/Ziad_Rafiq_Beydoun#/media/File:Don_Beydoun.jpg

Page 107: https://en.wikipedia.org/wiki/Harry_Hammond_Hess#/media/File:Hess.gif

Page 108 left: https://en.wikipedia.org/wiki/Bear_Seamount

Page 108 right: https://en.wikipedia.org/wiki/Seafloor_spreading#/media/File:Age_oceanic_lithosphere,_Muller_et_al.,_2008.jpg

Page 110: ©Halliburton

Page 111: ©Professor Fred Vine and reproduced with his permission.

Page 112 left: https://upload.wikimedia.org/wikipedia/commons/c/cc/East_Pacific_Rise_seafloor_magnetic_profile_-_observed_vs_calculated.png

Page 112 right: Redrawn from data in Heirtzler, J.R., Le Pichon, X. & Baron, J.G. 1966. Magnetic anomalies over the Reykjanes Ridge. Deep Sea Research and Oceanographic Abstracts, 13, 427-443.

Page 114: https://upload.wikimedia.org/wikipedia/commons/1/13/Geomagnetic_polarity_late_Cenozoic.svg

Page 116: https://commons.wikimedia.org/w/index.php?curid=43819116.

Page 117: ©Halliburton

Page 120 left: ©Archives Imperial College London and reproduced with their permission

Page 120 right: ©Archives Imperial College London and reproduced with their permission

Page 121: https://commons.wikimedia.org/wiki/File:Road_Cutting_-_geograph.org.uk_-_820828.jpg

Page 122: ©North-West Highlands Geopark and reused with their permission

Page 123 top: Courtesy of Dr Andrew Davies, Halliburton

Page 123 bottom: ©Sedgwick Museum of Earth Sciences and used with their permission

Page 124: ©Sedgwick Museum of Earth Sciences and used with their permission

Page 125: https://upload.wikimedia.org/wikipedia/commons/0/0e/Diamictite_%28metatillite%29_%28Konnarock_Formation%2C_Neoproterozoic%2C_~750_Ma%3B_Rt._603_roadcut_near_Konnarock%2C_Virginia%2C_USA%29_22_%2822794318668%29.jpg

Page 126: ©The Geological Society and reproduced with their permission

Page 127: https://upload.wikimedia.org/wikipedia/commons/8/8a/Plates_tect2_en.svg

Page 129: ©Northwestern University and used with their permission

Page 130: AAPG ©1987, reprinted by permission of the AAPG whose permission is required for further use.

Page 131: AAPG ©1977, reprinted by permission of the AAPG whose permission is required for further use.

Page 132: https://www.nasa.gov/multimedia/imagegallery/image_feature_152.html

Page 133: top: ©Prof. Maureen Raymo and used with her permission

Page 133: bottom https://en.wikipedia.org/wiki/Wollaston_Medal#/

Page 134: https://upload.wikimedia.org/wikipedia/commons/6/60/Five_Myr_Climate_Change.png

Prof. Mike Simmons

HALLIBURTON TECHNOLOGY FELLOW

Mike is responsible for the investigation into innovation in geoscience as applied to the hydrocarbon exploration process. Previously, he was Earth Model Director at Neftex and before that worked at BP, Aberdeen University and CASP at Cambridge University. His main interests are applied stratigraphy and the geology of the Tethyan region. He is a Scientific Associate of The Natural History Museum in London and Honorary Professor at Royal Holloway, University of London.

(mike.simmons@halliburton.com)